A Geography of Digestion

CALIFORNIA STUDIES IN FOOD AND CULTURE

Darra Goldstein, Editor

A Geography of Digestion

BIOTECHNOLOGY AND THE KELLOGG CEREAL ENTERPRISE

Nicholas Bauch

UNIVERSITY OF CALIFORNIA PRESS

University of California Press, one of the most distinguished university presses in the United States, enriches lives around the world by advancing scholarship in the humanities, social sciences, and natural sciences. Its activities are supported by the UC Press Foundation and by philanthropic contributions from individuals and institutions. For more information, visit www.ucpress.edu.

University of California Press
Oakland, California

Library of Congress Cataloging-in-Publication Data

Names: Bauch, Nicholas, author.
Title: A geography of digestion : biotechnology and the Kellogg cereal enterprise / Nicholas Bauch.
Other titles: California studies in food and culture ; 62.
Description: Oakland, California : University of California Press, [2016] | Series: California studies in food and culture ; 62 | Includes bibliographical references and index.
Identifiers: LCCN 2016031590 (print) | LCCN 2016033511 (ebook) | ISBN 9780520285798 (cloth : alk. paper) | ISBN 9780520285804 (pbk. : alk. paper) | ISBN 9780520961180 (Epub)
Subjects: LCSH: Kellogg Company. | Battle Creek Sanitarium (Battle Creek, Mich.) | Digestion—Environmental aspects—United States. | Breakfast cereals—United States. | Sanitary engineering—United States—History. | Cereal products industry—Technological innovations—United States.
Classification: LCC HD9056.U6 K4535 2016 (print) | LCC HD9056. u6 (ebook) | DDC 338.7/6647560973—dc23
LC record available at https://lccn.loc.gov/2016031590

Manufactured in the United States of America

25 24 23 22 21 20 19 18 17
10 9 8 7 6 5 4 3 2 1

The publisher gratefully acknowledges the generous support of the Director's Circle of the University of California Press Foundation, whose members are:

Harriett & Richard Gold
Gary & Cary Hart
Robert J. Nelson & Monica C. Heredia
Marilyn Lee & Harvey Schneider
Thomas & Barbara Metcalf
Jerome & Ann Moss
Barbara Z. Otto
Margaret L. Pillsbury
Larry & Rosalie Vanderhoef

To Yi-Fu Tuan

Contents

List of Illustrations ix
Acknowledgments xi
Introduction: Spatially Extending the Digestive System 1

1. The Battle Creek Sanitarium: A Place of Health 20
2. Scientific Eating: Kellogg's Philosophy of the Modern Stomach 46
3. Flaked Cereal: The Moment of Invention 77
4. Extending the Digestive System into the Urban Landscape 102
5. The Systematization of Agriculture 124
6. Breakfast Cereal in the Twentieth Century 155

Epilogue 172
Notes 175
Bibliography 201
Index 215

Illustrations

FIGURES

1. Electric bed 41
2. Examining office 50
3. Stomach tube 50
4. Digestion chart—stomach pump results 51
5. Laboratory men at work 52
6. Analysis of stomach fluid chart 56
7. Lab report 58
8. Chemical exam chart 59
9. Diet chart 65
10. Sanitarium foodstuffs 67
11. Granola advertisement 85
12. William Kellogg scraping flakes of wheat from the rollers 91
13. Granose advertisement 94
14. Sewer map application 105

15. Stomach gates 117
16. Daily schedule of food movement through digestive tract 118
17. Colonic machine 120
18. Map of Goguac Lake 121
19. Close-up of colonic machine 122
20. Stomach analysis results 130
21. Threshing machine 146
22. Vibrating plates 146
23. Hammond Seedsman seed-shipping department 151
24. Inventory of food quantities 151
25. Hammond Seedsman mail-order department 152
26. Hammond Seedsman manufacturing plant 153

TABLE

1. Daily calories table 139

Acknowledgments

At its publication date, this project is eleven years old. I use the term *project* with intention, because while its current state is a book, the book itself feels like a shoot of an underground rhizome that has pierced the soil and made it to light. The underground rhizome is the ever-growing, twisting and turning phenomenon that we might call an intellectual journey, or maybe even a career. Like many scholars, feeding this rhizome from time to time I find myself using the online *Oxford English Dictionary* in search for inspiration, which is exactly what I was doing when I discovered an obsolete (Shakespearean, actually) usage of the term *project:* "a mental conception, idea, or notion." In this way, *A Geography of Digestion* is a project in the sense that it began as a vision, something that I saw clearly in my mind and then spent a long time figuring out how to articulate. It is also a project in that it is ongoing, never fully complete, and open to conversation and growth; it may always feel notional.

I would like to thank the people directly involved in this particular shoot of the rhizome, that is, the making of the book. I respect and value the anonymous peer-review process, and this book would not be what it is without two extremely patient, thoughtful, and giving reviewers. It is an honor to have peers so carefully engage with one's core thoughts and styles of articulation.

The editorial staff at the University of California Press remained committed and involved in the making of the book. Thank you to Darra Goldstein for inviting me to submit a "rhizome shoot" (my words) to be considered for inclusion in the California Studies in Food and Culture series. Many of the scholars I admire most in food and agriculture studies have contributed to this series over the past fifteen-plus years, and it is humbling to have this project find light here. At the press's offices it has been a pleasure to work directly with Stacy Eisenstark, Zuha Khan, Glynnis Koike, Bradley Depew, Tom Sullivan, and Dore Brown. Kate Marshall took the time to meet with me in person, shepherding, shaving, and refining the content to help transform it from an idea to a book. Thank you to Emily Park, who copyedited every character of the manuscript with profound consistency and astuteness. And thank you to Carol Roberts, who took the time to read both the surface and depth of the text while crafting the book's index.

I was able to perform archival research throughout the state of Michigan thanks to the guidance of many librarians and keepers of history. From Michigan State University thanks go to Anita Ezzo, Cynthia Ghering, and Kathleen Weessies. From the University of Michigan, thank you to Karen Jania. In the town of Battle Creek itself, where all the local geographical knowledge exists, thanks especially to Garth "Duff" Stoltz, Andrew Michalowski, and George Livingston. Not all of the Kellogg manuscripts are in Michigan; many of them are carefully housed in Loma Linda, California, one of the present-day headquarters of the Seventh-day Adventist denomination. At Loma Linda University, thank you to Lori Curtis and Janice Little. Generous funding for carrying out this research was granted to me through travel and research fellowships from the American Association of Geographers' Cultural Geography Specialty Group, the UCLA Darling Biomedical Library, and the University of Michigan's Bentley Historical Library.

I wrote this book—or rather rewrote it based on my dissertation work—while a post-doctoral researcher at Stanford University. Thanks go to Zephyr Frank, who, despite my simultaneously conducting another entirely different major project, encouraged me to keep writing *A Geography of Digestion*. Moving between projects over my years at the Center for Spatial and Textual Analysis worked well for me, as the ideas always seemed to stay fresher that way.

So much for the book project itself, with sincere apologies to all those left unnamed. What about the acknowledgments one makes to the rhizome, or the greater building of one's intellectual identity? Here I could thank hundreds if not thousands of people. My interest in food and agriculture studies began in the University of Wisconsin-Madison's geography department, which has a long tradition of research in nature-society relations. During my time as a graduate student at UCLA, eleven years ago, the people who initially helped get this project going include my main advisor, Michael Curry, who was always wise and patient in the evolution of the project; he remains someone I look up to with great admiration. Denis Cosgrove, along with many other students, taught me the power of landscape as an analytic tool. John Agnew helped frame the broader political significance of food and eating. And Norton Wise, in whose history of science seminar I discovered that John Kellogg cared about digestion, offered me an inroad to write and talk about the materiality of food consumption. During my first year at UCLA I had the opportunity to participate in a University of California intercampus exchange program, which brought me to UC Santa Cruz for one spring term. There I had the privilege to meet regularly with scholars in their food studies program, including Julie Guthman, Melanie DuPuis, Margaret Fitzsimmons, and Melissa Caldwell, all of whom were generous with their time despite busy teaching and research schedules. I greatly value all of these experiences.

To my wife, Brenda: thank you for keeping the project light and fun. Though I do not eat cereal, she tells me it's delicious.

N.B.
Palo Alto, California
July 26, 2016

Introduction

SPATIALLY EXTENDING THE DIGESTIVE SYSTEM

In 1896, exactly a decade before William Keith Kellogg put his famous signature on the first Corn Flakes box, his older brother, Dr. John Harvey Kellogg, published a book called *The Stomach: Its Disorders and How to Cure Them.* In it he outlined his philosophy of nutritional health as practiced at the well-known Michigan retreat-clinic, the Battle Creek Sanitarium. This philosophy rested on a digestive reductionism, presenting ailments and diseases of all kinds as curable through the proper consumption and digestion of foods. Helping patients digest food quickly, cleanly, and accurately was the foundation for healthful living at the sanitarium. What are thought of today as staples of American health foods—soy milk, peanut butter, granola, yogurt, and of course Corn Flakes—were all either invented or perfected for mass consumption in the experiment kitchens of the Battle Creek Sanitarium. The sanitarium was, in Kellogg's words, "a university of health." It offered organized exercise regimens, an ongoing lecture series, and a rather odd battery of machinery including electric light baths, automatic massagers, and abdominal punchers that were meant to improve health. But above all, John Kellogg's message of health in the decades preceding the mass production and exportation of breakfast cereal was centered on digesting the right foods in the right way.

Digesting at the sanitarium, however, did not occur as a phenomenon confined to the insides of patients' bodies. It was, on the contrary, a very spatial—very geographical—process. When Kellogg centered health on the digestive system it meant, paradoxically, that the organs themselves had to be extended beyond the form of the body. Digestion could not happen without the aid of extracorporeal technologies, what today we might see as older types of biotechnologies. These late nineteenth-century biotechnologies came in the form of food-processing machinery, architectural environmental systems, urban sewerage, and agricultural implements for the mechanization of making food. Each of these biotechnologies happened in particular places. Fleshing out rich descriptions of those places shows us John Kellogg's geography of digestion, while at the same time demonstrating an instance of what literary theorist Stacy Alaimo has recently called "a sense of the subject as already part of the substances, systems, and becomings of the world."[1] It is important to recognize that "subjects"—sanitarium patients, in this case—are ever involved in greater environmental and technological systems. This book aims to demonstrate that being a biological human neither is nor has ever been achieved without an ensemble of other "substances" that can be mapped onto the surface of the earth.

A Geography of Digestion, therefore, outlines one significant origin of American health food–ism from the perspective of the body, and specifically the digestive system. With the meteoric rise of popular consciousness concerning the geographical origins of food in the twenty-first century, this book demonstrates another way to think spatially when we talk about food. Countless maps have been drawn showing where food is grown, how far it travels, or the labor injustices associated with growing food in distant places. But these discussions tend to stop when food reaches the proverbial kitchen, or plate. Though the topic of digestion has been approached more recently by food scholars, there remains an outstanding question concerning the geography of food once it has been ingested.[2]

To forge this other way of thinking about food spatially, the book connects the body with the landscapes, machinery, and urban infrastructures that emerged alongside John Kellogg's invention of the "modern stomach" in the 1890s, an invention that reflected state-of-the-art advances in nutritional and medical science. Yet, far from a set of organs confined within the epidermal bounds of the body, the digestive system existed in other places. Without recognizing this we miss a major opportunity to clarify the particular relationship between food, body, and environment at a crucial moment in the emergence of American health food sensibilities.

Stepping back a little further, the book grapples with the question of how we are to understand the relationship between body and environment, an issue of increasing concern within the environmental humanities and health geography. Asking "where is the digestive system?" allows me to spell out in detail how environmental thinkers might start describing the thought that Alaimo captures in her notion that subjects are *already* part of the substances of the world. This book shows that describing body–environment relationships depends on the ability to create a clear picture of the tools and technologies available to the people living in that context.

Bodies at the sanitarium could have neither functioned nor healed as they did without the aforementioned technologies, all of which existed within the milieu of southern Michigan. All of these technologies became part of patients' bodies as soon as they stepped off the train in Battle Creek and took their first bite of health food. This suggests that with careful study one can read the meaning and makeup of bodies from the landscapes that constitute them; interpreting landscapes tells us about our material selves as much as it does about the values and actions of a particular culture. The politics of this vision are, again, geographical. That is, in a concept of reality—an ontology—defined by the relationship between bodies and technologies, policy makers should be open to solutions that exist in places other than ground zero, so to speak, of any particular problem. When the solutions to health epidemics focus only on bodies, for example, it is possible that an entire map of potential places on which attention could be focused is lost.

John Kellogg wrote in his definitive book on the subject of digestion, "it may truly be said that disorders of digestion are the most prevalent of all human ills . . . the great majority of diseases are primarily due to derangement of the digestive processes."[3] The stomach, therefore, was the focal point for Kellogg's brand of healing at the sanitarium, much more consistently than the quirky experiments—like electricity baths or punching machines—for which he is sometimes known. But if we allow ourselves to dive into Kellogg's world and his logic, it is apparent that the stomach could not have acted alone in bringing his patients back to health. For this we must look to the local landscapes in which digesting food took place.

In the ever-evolving conversation about eating healthfully in American culture, digestion has, and often still does, serve as a diagnostic tool for the quality and healthfulness of foods. Bad food gives stomachaches, indigestion, or, more seriously, Crohn's disease. Good food, on the other hand, quietly disappears into the fibers of the body without a fuss. Indigestion is such a common concern surrounding well-being that medical practitioners have discussed it for centuries. Digestion has remained a focal point for achieving health through the present, as well, with sustained medical attention on illnesses such as irritable bowel syndrome, or the procedure of fecal bacteriotherapy, in which healthy fecal bacteria are transplanted into a sick patient's digestive system.[4] And yet the process of digesting food has only recently come under the purview of scholars in food studies.[5] Digestion is so hidden, so seemingly contained within the body, that it begs to be categorized as a discrete process. But there is an environmental geography for digestion, too, just as there is a spatial-material story for any phenomenon.

Exploring this case from the vaults of American eating shows how technologies define the processes and the *locations* of digestion. Kellogg used steel dough rollers to invent flaked cereal, a food item that he thought was more readily assimilated into the body, thereby relieving the stomach from working too hard. The rollers were the first consciously applied prosthetic for the digestive system in American cuisine, the first intentional geographical removal of an alimentary organ function to another location. Kellogg sat on the Michigan state board of health for over ten years, helping usher in the town's sewage system, which altered the urban geography of digestion for his patients. In addition to these two instances, digesting food cannot be fully portrayed in this time and place without including the origins of the foods themselves, the agricultural economics that made the nationwide manufacture of breakfast cereals possible.

The invention of flaked cereal by the Kelloggs accelerated the deliberate technological intervention in the making of healthy bodies in Battle Creek and the nation at large. Based on John Kellogg's lifelong, if troubled, affiliation with medical science communities, and his obsession with digesting food "properly," flaked cereal was seen as the antidote to numerous ailments held by the sanitarium's clientele. Kellogg designed flaked cereal with special precision so as to maximize the efficiency of digestion while minimizing discomfort and the debilitating effects of eating that were common in the nineteenth century. Problems with digestion in this period often resulted from eating food in which lived an overabundance of harmful bacteria, resulting in dyspepsia (a generic term for bad indigestion) at all levels of severity. Kellogg saw flaked cereals as a greater achievement than any food eaten directly from nature. But technological intervention in the manufacture of foods, and the entire process of digesting them, meant that the technologies themselves had to exist somewhere.

LANDSCAPE AS EPISTEMOLOGY

Machines outside our bodies act as extensions of the organs inside our bodies. Going beyond the *idea* of bodies existing in relationship with other technologies, how exactly *are* bodies extended into the world? Practicing health care must in part be geographical because under an ontology of spatial extensibility, health can be read from landscapes as much as it can be diagnosed from peering into bodies.

Bringing digestion outside the body—following what were really biotechnologies—means that a different set of research techniques are employed than would normally be when studying a biological process. In particular, two core concepts from cultural geography—landscape and network—guide the interpretation of the organs of digestion (including the stomach, intestines, and colon). The point of the project is to explicate how three technologies developed in southern Michigan—"pre-digesting" cereal rollers at the sanitarium, urban sewage infrastructure in Battle Creek, and agricultural machinery in the town's hinterlands—were performing digestion for sanitarium patients. The small, unassuming town of Battle Creek becomes a perfect picture of the ways in which food, technology, landscape, and the making of healthy bodies came together in American history. This explication opens up a methodology for representing bodies that are materially and spatially extended beyond the epidermal borders of the skin.

It is important to recognize that even in the controlled atmosphere of a late nineteenth-century health resort, entire urban and rural landscapes full of rich, intertwined connections among people, objects, and ideas forged something as visually hidden and as taken for granted as digesting food. In building this geography, the abdomen is figuratively incised, unveiling a biological process that can be observed with the naked eye if one is in tune with the landscapes to which it is connected. Landscape in this way becomes an epistemology—or a way of knowing—that explains phenomena from the perspective of the terrestrial and the technological. As much as *actually* cutting the skin and peering into the digestive system, we can learn about the hidden body by studying what it is connected with, or what sustains it.

Instead of seeing landscapes as material objects of study, we can see them as the gateways, or translators, to understanding other phenomena that may not appear to be associated with the landscape itself. Using landscape as an epistemology allows me to build up the idea of the extensible body—that is, a body that cannot be explained without understanding the landscape. One can, therefore—ideally, at least—read the digestive organs from the landscape. And more generally—crucial for geographical practice—one can start to understand a thing by looking *around* it, rather than *at* it. This means that as researchers and as curious people, we can come to know what appear to be singular objects by studying broader-scaled landscapes, the spaces in which we move around and touch. Landscape-as-epistemology brings the human experience in tune with the everyday scene.

The methodological move is to *begin* with the landscape, assuming that reading it reveals much about the object of interest. This method is vastly different from one that begins with the digestive system. Starting there would leave a researcher focused on volumes of scientific publications that debate the best practices of digesting food and treating digestive disorders. The researcher would hone in on the minutiae of bacteria, acids, and blockages, referencing other body systems but seldom lifting his head to take a wider view outside. That the surrounding local landscapes might play a role in the process of digestion, and even in the *making* of the organs, demands—from the perspective of medical science—a radically different approach. Beginning with the landscape to investigate the digestive system assumes from the start that *where* this particular, historically significant style of eating happened had a lot to do with *how* it happened. To embark on this historical-geographical investigation with a view already backed away from the body means that this book contributes to discussions going on in a variety of academic corners, including food studies, cultural geography, network theory, new materialism, and object-oriented philosophy.

To see how this methodological approach is useful, consider other fields of environmental studies. For example, work in environmental history and environmental justice shows that the health of landscapes is connected to the health of bodies. In this vein, scholars have used toxicity as a framework to demonstrate the ill effects that polluted landscapes have on the health of disenfranchised populations.[6] Poisonous industrial waste, in this way, is used by environmental health scholars not unlike the way barium is used by present-day health-care professionals. To explain briefly: in a common method for tracking the movement of material as it courses through the body, doctors will sometimes instruct patients to swallow a liquid containing barium, which is then observed via x-rays. In this diagnostic procedure, the dispersal of a single element—barium—is visualized as a proxy for how the body circulates material. Just as barium uncovers metabolic processes at the scale of the body, scholars in environmental studies have used toxic quantities of elements—including diethylstilbestrol (a synthetic estrogen) and organochloride pesticides (e.g., DDT), among others—to track the metabolism of these poisonous chemicals as they course through the scale of the landscape and, of course, people's bodies. Rachel Carson's *Silent Spring* is foundational in this type of environmentalist thinking, from which there has developed a strong sense that the environment *as a thing* matters because it affects people and other living organisms. The pump is primed, therefore, to develop methods by which those effects can be described. But the newness of this method is that the effects must be described in an intellectual ferment where the very categories of nature, human, and artifice have merged, where "environment" and "technology" are not singular things, but already part and parcel of being human.

WHY DIGESTION?

The material meeting point of the outside world with the human body is the digestive system. Critical observers of agriculture will know that from the seed itself, as edible plants emerge from the soil they are brought into a system designed by people, and the effects of those designs are legible through their journey "from farm to table" and all the way to the inside of the body.[7] The digestive system, therefore, is a critical nexus on which to focus—one that promises to efface the boundary between body and landscape. Eroding the division between inside and outside suggests that landscape *is*, or at least under certain circumstances *can become* the body itself, and that when we utilize technologies to remake landscapes we are undoubtedly remaking populations of people as well.[8]

Scholarship in food studies can be divided into two major camps. The first are works on food production, which have tended to focus on agricultural technology, farm labor, political and economic analyses, commodity chains, and alternative agriculture. The second are works that explore food consumption, the emphases of which have been more on culture, race politics, affect, taste, health, and anxiety. The former camp has a particularly material bent to it, using concepts of landscape and environment as pillars in its approaches, while the latter group looks to social theories to understand why people eat what they do, and what it means in a greater political context. Digestion brings these two schools together.[9]

The implicit and powerful insight from the work of environmental scholars who study toxicity and health is that the landscape–body relationship is a fundamentally material one, where danger comes in the form of (frequently invisible) objects. Topically, toxicity plays only a supporting role in *A Geography of Digestion*, and yet the project nonetheless explicitly furthers the toxic-environment research agenda by describing the mechanics of how technologically modified landscapes become the body. To do so requires—as has just been hinted at—an "object-oriented" approach, where the spatial boundaries of objecthood are taken under investigation. This method—a blend of regional geography, actor-network theory, and object-oriented philosophy—clarifies what the objects being connected are, but it also goes further, theorizing the very constitution of those connections.[10] By doing so, my method in this book gives theoretical and practical juice to the present environmentalist agenda. To put it simply, we know that the objects (e.g., toxins) that exist in an environment affect other objects (e.g., people) that are nearby, and yet what lacks is a way to describe how those objects become materially connected in space. One major aim of *A Geography of Digestion* is to show an example of how this might be accomplished using historical-geographical description.[11]

This re-presentation of health food's past points toward a new method that uses the relational materiality of actor-network theory in the cradle of regional, geographical description. Actor-network theory has brought into focus the idea that phenomena (e.g., digesting food) are multi-object, and multiply placed, events. Geographical description brings a subjective, artistic practice to bear on communicating what those multiply placed phenomena look like. The melding of these two approaches yields an important method because it actualizes ideas about how networks function, a topic of great interest to political ecologists, environmentalists, and spatial practitioners concerned with materiality. The theoretical conclusion is that by using landscape as an epistemology—that is, by using landscape as a way of knowing how digestion works—then the way objects (like stomachs and sewers) are *related* becomes instead an issue of how one is an *extension* of the other.

A Geography of Digestion takes a case from the past that, on the surface, has nothing to do with biotechnology, landscape, or even digestion, and has everything to do with American cuisine, health food, and a moment in the small-town, nineteenth-century Midwest that spurred a revolution in eating. The work remains equally focused on the locale of Battle Creek, using its urban and rural landscapes as drivers in the building, or assembling, of this very spatial, very material network of digesting food. As this approach facilitates an investigation into what it means for entities to be "related," my aim is to wield this method to build a way of thinking such that object connectivity is a given.

JOHN KELLOGG AND BATTLE CREEK

In his long career, John Kellogg published hundreds of articles, pamphlets, and books, summoning a tidal wave of ideas, propositions, hypotheses, conclusions, statements, and advisories. His prolificacy leaves plenty of room for variation in the quality and cohesiveness of the oeuvre. Reading Kellogg long enough, one is bound to find competing fact claims at one point or another. Sometimes these discrepancies have a logical historical progression to them, while at other times Kellogg appears to be saying whatever makes him sound authoritative in the context of the conversation or his present audience. Reading the lifework of someone who incited a lasting global corporation by ceaselessly operating at the highest level of productivity, it would be impossible to represent everything that Kellogg wrote throughout his life in one book. However, in going through so many of his publications, it is clear that a lasting tenet in Kellogg's overarching goal of promoting health was about designing foods and designing bodies so that they fit together seamlessly.

Unlike many of Kellogg's contemporary health reformers, Kellogg did not believe that human bodies had ever had a pure relationship with food. Rather, in his estimation, that relationship was broken from the start. Modernity was not creating the problem for Kellogg—with its processed flours or packaged breads—but rather is where he found the solution. Food and bodies could both finally be made in a way such that they complemented each other, solving sickness, degeneration, and disease. This was squarely, as Foucault would have it, a biopolitical project.[12] By breaking apart and categorizing the process of digestion into pieces, then outsourcing those pieces to various infrastructures and technologies at different geographic scales, Kellogg aided in the control of people's bodies by the institutions that managed the technologies. In some cases, the Kellogg corporation itself changed the course of what it meant to be healthy, while in other cases the urban, state, and even federal government's policies on sewerage, land tenure, and nutrition were responsible.[13]

The first part of this book is a history of John Kellogg and the Battle Creek Sanitarium, focusing on his scientific approaches to eating and digestion.[14] The second part of this book moves through the landscapes in which Battle Creek was situated. The traverse is scalar in nature, meaning that it starts closest to the modern stomach itself and moves outward, encompassing first the city of Battle Creek and then the agricultural hinterlands of southern Michigan. This spatial narrative draws a clearer portrait of what had to happen away from the dinner table, and outside the patient's direct experience at the sanitarium, in order to make Kellogg's method of eating and digesting food possible. This approach assumes nothing about the appearance of foods in front of patients. Where was the food grown? How did it arrive at the sanitarium? What were the conditions in the fields that enabled the Kelloggs to acquire foods at an affordable price? Further, the approach assumes nothing about where the food went after it was digested. Healthy digestion for Kellogg did, after all, require the frequent expulsion of waste from the body. If we are concerned with the implications of studying digestion within food studies, then we must also consider the pathway of human waste—with all of its public health and built environment implications. By carefully studying how food was made, eaten, digested, and disposed of, the categories of inside-body and outside-body break down, giving way to a space produced by the extension of objects to one another.

FOOD AS MEDICINE

Kellogg's new system of eating was determined by his own scientific research on the efficiency and nutritional benefits of a wide range of vegetarian foodstuffs. His laboratory program emerged alongside the work of the government nutrition researcher Wilbur Atwater, whose dietary charts Kellogg emulated in the sanitarium's publications.[15] Beginning with a deep understanding of the anatomy of the digestive organs that he gained from medical school training in the 1870s, Kellogg spent most of his career attempting to advance an ever-more-nuanced gastroenterological knowledge base. How the body took on the beneficial elements of food, and how it eliminated waste, were profound markers of one's overall health. The contemporaneous development of the science of eating and the science of digestion resulted in a philosophy of health wherein Kellogg used food as a medicine to cure ailments across the board, from cholera to skin rashes to depression.[16] As his research findings grew in number and complexity concerning the relationship between food and the bodies of his patients, though, he increasingly expected more from the food items. By the early 1890s—about fifteen years after Kellogg assumed directorship of the Battle Creek Sanitarium—there was no food from nature that could affect the body as precisely and as effectively as could the foods that he himself crafted.

In a move that would introduce a new way of eating to the American public that was reproduced with little restraint until the 1960s, Kellogg sought to replace nature with technologies. The pattern of packaged, mass-produced food that Kellogg started—still today widely available in supermarkets—was marketed as safe, nutritious, non-perishable, and tasty. Though packaged and preserved food is commonly associated with mid-twentieth-century American cuisine, it finds some of its strongest roots in Kellogg's philosophy and business practice.[17] To make foods that most perfectly fit how the body's digestive system functioned, and therefore most perfectly cured his patients, meant that Kellogg's pharmacological use of foods needed a pharmaceutical company, as it were, to make the drugs. The interventions Kellogg made with food-making machinery resulted in designer foods, yielding what in his system of health were antidotes to putrefaction and auto-intoxication, the root causes for a host of crippling diseases that began in the stomach.

Now, over a century after Kellogg's prime, prevailing thought toward the digestive system's ideal state has undergone an about-face. Instead of the sterile, abacterial stomach that Kellogg prescribed around the turn of the twentieth century, there are now movements that attempt to cultivate bacteria in bodies through probiotic programs.[18] Kellogg popularized the idea that one's own food acquisition and cooking techniques could never eclipse the health benefits of foods that were manufactured under the banner of nutritional and digestive science. The positive public reception of this stance is demonstrated by the sudden proliferation of health food companies in Battle Creek and beyond in the late 1890s, as a copycat economy sought to fulfill the demand for foods manufactured with a specific purpose for generating greater health.[19]

The technologies that Kellogg used to remake the relationship between bodies and food inflected what American cuisine would become in the first part of the twentieth century. These were foodstuffs that had been fortified with the knowledge of chemistry. These were the items in the modern American cuisine that could not escape the gaze or the meddling hand of the technician, who proceeded to redirect nature's edible bounty into something much more functional, safe, and nutritious.[20] This is the cuisine against which its antithesis—organic farming—has now popularly railed for decades. The ethos of organics has been to stop the meddling, measuring, and analyzing, instead practicing a trust in foods that are imbued with the irreproducible power of nature rather than the fleeting folly of human knowledge. But the strength of this reaction demonstrates the overwhelming strength of previously prevailing economies and cultural attitudes toward food safety and nutrition that were born from technical interventions like Kellogg's.

TECHNOLOGIES OF DIGESTION

Three technologies shaped the digestive system, and therefore health, for Kellogg. The first technology was a series of industrial kitchen tools used to make flaked cereals. The Kellogg brothers heralded their cereal-flaking process as one that would yield the single food product most prepared for digestion. And the tool that most singly accomplished this was a set of 8-by-24-inch steel rollers. Used originally to crumble large, dry pieces of baked granola into bite-sized pieces, the Kelloggs repurposed the rollers as flatteners of moist, dough-like globs of wheat mush. This flattening process, once perfected, would make flakes that could then be baked in a way that most effectively converted their natural starches into dextrin. Up to this point in 1894, the Kelloggs had already made dozens of baked food products, always believing that high oven heat healthfully transformed grains into sterile food suitable for human consumption. But the shape of the grains being baked—the flake—is what made possible the most complete conversion of starch to dextrin, and therefore the food most prepared to "prompt assimilation," as John Kellogg put it in an 1894 advertisement for wheat flakes.[21] To Kellogg this meant that the body's digestive organs could do the job he thought they were supposed to do—that is, to find and incorporate the valuable elements of foods while eliminating the nonvaluable ones—without the grinding, painful, and even harmful duty imposed on them by imperfectly conceived and engineered foods. The rollers, therefore, are one instance of the outsourcing of the toil of the digestive system to a place outside the body. The mechanical work of the rollers relieved pressure from the body.

For Kellogg, it seems that this was a modern technology solving the age-old problem of indigestion and its cascading ill effects. Equally plausible, however, is that it was a modern technological solution in search of a problem. In this case the invention of a food product, an object made for alimentary consumption but one that existed outside the body, needed a problem that existed inside the body. The search for a problem to solve brought Kellogg to the conclusion that his own culinary-technological tinkering had outpaced the digestive system. To bring the digestive system into the same league of progress represented by his foods, Kellogg needed to join the two; he needed to bring the outside in conjunction with the inside so that his technologically advanced foodstuffs had a problem to fix.

The roller machines—as well as the ovens, baking trays, and other kitchen implements—were originally housed in the experimental kitchen of the sanitarium; then, after production increased, they were moved to the Sanitas Food Company's factory in Battle Creek. The second technological intervention in the process of digestion—the erection of the town's new sewer system—though, happened somewhere else. If chapter 3 is about the material making of efficiency in the body's digestive organs, then chapter 4 is about the material effects of that efficiency. Kellogg promoted a thrice-daily evacuation schedule for his patients, such that each meal prompted a clocked turnaround from food to feces. The great, underground sewers in the largest European cities began functioning as early as the 1850s.[22] But selling such a dramatic revolution in public health, especially across continents, was not always easy. And so it was not until the late 1880s that the southern Michigan town of Battle Creek began implementing its own sewage infrastructure, relying before then on cement vaults, privies, bedpans, and outhouses. Great debates raged at this time, not only in Battle Creek but throughout towns in Michigan and the United States, over issues such as disease vectors, the role of soil as a filter, and the power of noxious odors. Whether you believed sewage was etiologically dangerous because of its foul-smelling fumes or because of microscopic life-forms called bacteria, urban populations increasingly sought to remove themselves from their own waste as the turn of the century approached.[23]

The most cogent salesmen for this expensive infrastructural revolution, though, were the ones who could show constituencies that sewerage was filled with harmful life-forms that, if ingested, were directly responsible for myriad diseases such as cholera, dysentery, and typhoid fever. And among the most cogent of these salesmen in all of Michigan was none other than John Kellogg. Kellogg brought his medical training to bear on the microscopic analysis of human waste. If food contained bacteria such that it needed to be blasted with oven heat to sterilize it, then human feces were far worse. In the 1890s all bacteria was bad bacteria for Kellogg, and he spent a great amount of energy showing this to Michigan's state board of health, where he was an active council member for over ten years. Kellogg's logic of efficient movement of food through the body to avoid self-, or auto-intoxication, as it was called, was mirrored in his support of constructing water-carriage underground sewers in Battle Creek. The efficient removal of waste as it coursed through the urban landscape was as hidden as the movement of food through the body. In chapter 4, then, the sewer system is investigated as the next phase in the geography of digestion. Sewerage was the material consequence of a bodily process that was outsourced to an urban landscape. As with the rollers, this technology can be read as an extension of the digestive process.

While John Kellogg played a part in bringing the sewage system to Battle Creek, he was involved less directly in that technology than he was with the grain rollers. Many others—including engineers, construction workers, and public officials—are players in the story of sewerage as well. As we move farther and farther away in space from the bubble of the sanitarium, Kellogg's immediate role continues to decrease. This is not to say that the third and final technology had less to do with how digestion functioned in Battle Creek. Rather, it is to point out that as we follow a transect away from the sanitarium, through the town and into its agricultural hinterlands, it becomes clearer why this book is best conceived as a *geography* of digestion. Maintaining a focus on digestion as a spatialized biotechnology is a matter of following the technologies that affected the body into the surrounding landscapes.

This suggests that landscape is the geographical tool that, when put into conversation with network, becomes an epistemology of the digestive system. And so, in chapter 5, the machines of agriculture are investigated as biotechnologies that—whether Kellogg recognized it or not—affected how digestion happened at the sanitarium. The chapter shows that two seemingly disparate spheres of society—the pharmacological consumption of food by American bodies, and the mass production of crops on American farmlands—were undergirded by the same intellectual thrust of the time that looked to chemistry for a way to maximize efficiency and production, both for bodies and for land. One outcome of this chemical knowledge was the production of food that was further transformed by cultural interpretation (Kellogg's, in this case) and turned into health food. The crucial place where agricultural science and nutritional science met was the stomach. In chapter 5 we see how midwestern agriculture in the late nineteenth century was becoming increasingly mechanized, reliant on science, and large scale, able to produce the grains en masse that would cure the ailing modern stomach as prescribed by John Kellogg at the Battle Creek Sanitarium. Grain-threshing machines form the basis of the conversation in chapter 5, as two agricultural manufacturing companies operating out of Battle Creek—Advance Thresher and Nichols & Shepard—designed farm tools that were shipped around the Midwest and the world. To cure the stomachs of people beyond the sanitarium, John and Will Kellogg needed access to an agricultural economy that supplied them with cheap grain. The transformation of the richest soil in the world into an agricultural behemoth happened with no small thanks to the machines engineered in Battle Creek.

And so we have a picture of an emerging network of objects that come together to create a singular phenomenon—the digestive system. Seeing the digestive system as a phenomenon implies that it is a happening, a coming together, rather than a singular item, or what we would usually call a body organ. In the context of this study, I prefer to call the digestive system a biotechnology because the constituent parts that comprise this "posthuman" organ have a distinct spatiality. This space, or geography, of the digestive system is defined by the bounds of a network, one that exists in local landscapes. Reading these local landscapes allows us to peer back in time and witness the digestive system. In her landmark introduction to posthumanism, literary critic Katherine Hayles sees the "body as the original prosthesis we all learn to manipulate, so that extending or replacing the body with other prostheses becomes a continuation of a process that began before we were born."[24] This always-ever type of prosthesis is part of the theoretical foundation of *A Geography of Digestion*. In this book, though, the idea of a body stretching, or extending, into space is removed from the imagination and mapped onto the real landscapes surrounding Battle Creek. This is a geographical move that has not been thoroughly developed in posthumanism research.[25] Environmental-political theorists Diana Coole and Samantha Frost have pointed out helpful directions, too. In their essay on new materialisms, they ask what the place of embodied humans is within a material world in an era when the reprisal of materialism must be "radical."[26] *A Geography of Digestion* is a method and case study for addressing this question.

BETTER FOODS THAN NATURE

The main historical insight of this book is that through his calculated, rational way of making foods, John Kellogg popularized and nationally exported the concept that science could make better foods than nature. Furthermore, he exported the notion that the human body was not prepared to incorporate nature's bounty—vegetable or not—without precise technological intervention from experts in medicine, agriculture, public health, and nutrition.[27] In this view, American cuisine is defined less by a spread of foodstuffs and more by a relationship to food characterized by mistrust in the body's capacity to manage, distribute, and evacuate—in short, to digest—food in the most healthful way. That the food counterculture had something to react against is in large part, I believe, a result of Kellogg's mistrust in the body's nature.[28]

Food scholars have tended to overlook this point because of the association commonly made between Kellogg and the greater health reform movement, a Protestant-based, pseudoscientific social movement that promoted simple, bland, unadulterated tastes in food, sex, and dress. The health reform movement, that is, has been a red herring for interpreting the purpose of Kellogg's health foods. So, although he may have started as a foot soldier in the ranks of mid-nineteenth-century reformers—taking cues from the likes of health proselytizers Sylvester Graham and J. C. Jackson—Kellogg's relationship with the movement took a dramatic turn when he embedded himself deeply in the cause and methods of medical science. This commitment to the scientific understanding of digestion and nutrition, coupled with his rich background in Protestant health reform, meant that

Kellogg perpetually walked a tightrope between these two worlds. While the arc of his career (from the 1870s to the 1920s) trends away from the religious and toward the scientific, he never quite stopped leveraging this bimodal approach to appease as many patients as possible.

Because of this, I believe, Kellogg's legacy reflects the light of fame as much as it absorbs generations of ridicule. He alternated between research deemed progressive by the gastroenterological branch of the American Medical Association, and seemingly prank-like cures (e.g., electric water baths or abdomen punching machines) that irrecoverably twisted health reform philosophies in the popular imagination.

CONCLUSION

It is more accurate—as well as more environmentally informative—to know objects as processes involving many other objects. To ratchet up the potency of this claim—and to apply it to the phenomenon of eating food—is to acknowledge that "the digestive system" as such cannot exist, but that the thing we call the digestive system is the collapsing of myriad technologies that exist in a new type of spatial relation.

There are two paths one can follow in order to frame the geography of objects. One is to follow the spatial and historical threads of particular things, mapping out their unique geographies. This path brings objects in relation, or in association, with the industrial and labor processes that make their existence possible. These connections are frequently global in nature, spanning a variety of political-economic systems and social organizations of labor, as the material goods we consume travel around the globe to reach us. The other path we can take flips this strategy on its head, asking *not* "what is the cartographic representation of the life of an object?" but instead "what is the meaning of a place where an object ends up?" To use the strategy of association in this way requires us to ignore the impulse of following objects around, and instead to start seeing objects as they exist in relation to what is already around them.

Extensibility pushes the fundamental glue of actor-network theory—that is, relation—further away from the theoretical sphere and into the material-geographical one. When we see that a group of technologies—

food machinery, sewerage, and agricultural machinery—are external components of the digestive system, it opens a pathway for thinking about how exactly these digestive devices are connected. After all, they still appear to be distinct objects separated in space. Just because they affect one another, why should they *be* one another? The short answer is that the very idea of a discrete, essential object leaves us stuck in a world where phenomena cannot be understood as processes involving other things.

1 The Battle Creek Sanitarium

A PLACE OF HEALTH

In June 1863, not yet a month after the official inauguration of the Seventh-day Adventist religious sect, God spoke to its visionary leader, Ellen G. White. While at a friend's home in the southwestern Michigan town of Otsego, the "prophetess" lapsed into a dramatic trance, waking with a clear set of hygienic health laws. Though purportedly divine, the substance of her vision borrowed from the swirl of messages being pronounced by other mid-nineteenth-century health reformers: eat a vegetarian diet, avoid alcohol and tobacco, maintain personal hygiene, do not overindulge or overwork. In the case of sickness, medicinal drugs were to be avoided, replaced with natural and preventative solutions such as fresh air or hydrotherapy.[1]

These dictates, pillars of what historians generally call the American health reform movement, came from a variety of Protestant millennialist characters repulsed by the medical establishment, who believed fundamentally that maintaining one's corporeal health was a religious duty performed in anticipation of the second, judging, advent of Christ. Sylvester Graham, William Alcott, Joel Shew, Horace Mann, R.T. Trall, and James C. Jackson were the most influential health reformers surrounding Ellen White both before and after her 1863 vision.[2] While the contents of White's vision did not significantly diverge from that of the other health reformers, she captured the essence of the message in an energetic, compelling, and practicable way for thousands of people by giving health a place where it could be practiced with religion.

This chapter narrates the origins of the Seventh-day Adventist sect and the genealogy of the Battle Creek Sanitarium, as well as John Kellogg's food-based approach to medicine. Three topics arise in the course of this narrative: (1) the way that early Adventists understood bodies in relation to nature and medicine; (2) the importance of place in nineteenth-century healing practices, and in particular hydrotherapy's position as the predecessor to Kellogg's thinking; and (3) the seeming contradictions between science and nature (including the specific attributes of places) in Kellogg's ideas of health. The latter half of the nineteenth century was an era of great change in what it meant to be a professional physician. Together, White and Kellogg were a part of this change; studying their program of healing lends insight into the widespread uncertainty concerning how medical practitioners should adopt scientific insights about the body and disease. Kellogg found himself treading these uncertain waters frequently, adjusting his advice as his own understanding about digestion developed. A goal of this chapter is to link Kellogg's philosophy of digestion with the broader landscapes in which it developed—to describe, that is, how the landscapes of Battle Creek and its hinterlands became therapeutic ones, agents in the brand of healing undertaken by Kellogg at the sanitarium he invented.

Over the second half of the nineteenth century, the substance of White's vision would help generate an international following consisting of people seeking health and spirituality. Along with a prolific publishing house, the turn to health in the early years of the sect would prove its most successful strategy in concretizing a shared set of beliefs among its congregants, as well as its most lasting legacy. Notable food items popularized at the Battle Creek Sanitarium that are still considered staples of health include peanut butter, soy milk, and corn flakes. However, the development of the Seventh-day Adventist diet happened within the context of a broader concern for bodily well-being. Health became paramount to Ellen White and her adherents because it was a means of practicing their religion, a way to prepare for God's impending rapture. Because cultivating health was a set of actions determined by divine rule, the performance of health could be done together, in a communal, educational space. While one could spend a lifetime critiquing the veracity of Ellen White's visions, it is hard to dismiss the relevancy of her insistence on building an architectural manifestation of her health message. Just as a church is a building where people go to share spiritual practice—to make ideas real—White realized that a place of healing would galvanize her and others' enthusiasm for the ideas in her health message, just as the sect was beginning to grow large. Months after her vision on that day in June 1863, Ellen White wrote the following: "I saw that it was a sacred duty to attend to our health, and arouse others to their duty. . . . We have a duty to speak, to come out against intemperance of every kind—intemperance in working, in eating, in drinking, in drugging—and then point them to God's great medicine: water, pure soft water, for diseases, for health, for cleanliness, for luxury. . . . The more perfect our health, the more perfect will be our labor."[3]

These abstract ideas became dictates, and the dictates became practice. But the practice still needed a place in which it could be developed, performed, taught, discussed, and evangelized. White believed that if there was a place free from ridicule where people could congregate and observe how the health practices actually worked, then she would achieve her "duty" to speak out against intemperance of every kind. What would become by the late twentieth century a global pattern of Seventh-day Adventist health-care infrastructure started just down the road from Otsego, in the nearby town of Battle Creek, Michigan. At her home there, Ellen White forged ahead with the implementation of her plan to build a place where people would "learn to stay well."[4]

PLACES OF HEALING

What are the characteristics of a place of healing, where people learn to stay well? Spa towns are instructive in seeing how places curative or ameliorating to body and mind become produced. The modern construction of spa towns in Europe during the late eighteenth century and into the nineteenth followed a predictable pattern. A water or spring source was discovered in a rural or forested setting, which was then promoted by local business interests. A key step in the popularization of the water source was an accreditation by a chemist, followed by "a medical account of cures to be found at the place, preferably produced by a reputable physician." Promoters channeled these chemical and medical reports to elite communities, both local and distant, using them to differentiate among the varying curative properties found among the various spas.[5] Promoters, in essence, tried to match their places with a discrete set of illnesses, each spa offering a particular constellation of environmental factors promising to heal in specific ways. Though the granularity of distinction among spa towns never developed past a broad categorization of healing properties, the increasing success of spa towns is evidence of a clientele that valued the rhetoric of science—one that wanted to believe there was a place that could definitively make things better, and that experts had already done the work to prove it. The natural, curative properties held by spa towns, however, could not be exported. That is, as much as the waters of the spa were thought to heal, so too did the unique situated-ness of the town itself play a role in the process of getting well. The act of traveling to a distant, unpopulated setting—often in or near mountains, and often not easy to get to—has historically been a hallmark not only of spas, but also of other typical healing places.[6]

It is significant that what would become the Battle Creek Sanitarium—the place of healing in which John Kellogg would develop his brand of eating, digestion, and health—is a direct descendant of the European spa town. Healing at a spa required the corporeal co-presence of patient and place. Whatever ensemble of curative agents on which any one spa relied could be activated only on location. The thought of reproducing somewhere else the sometimes quasi, sometimes full-on spiritual quality of healing that transcended the sum of its parts was antithetical to the philosophy of going to the spa. That is, like most any travel experience, it was impossible to reproduce what it was like to be there.

Ironically, the logic used by chemists and physicians to convince would-be patients of a spa's reputability took an opposite tack. Whereas it became fashionable to travel to a remote place (the journey itself marking a commitment to dissolving disease) in order to gain the site-specific qualities of health, scientific explanations rested on the reproducibility of their results in other places. If the conditions of an experiment could not be re-created in another laboratory in another country by a different practitioner, then the rhetoric of science could neither predict nor speak with the tone of authority that the minerals found in spas would definitively change the physical state of one's body. Iron, for example, was iron, whether found in Michigan or Austria. To the chemist, geography was secondary, as minerals like iron, or saline, would affect people's health the same way whether found in a natural chalybeate or packaged and carried home to a city house and used there in the bathtub. If chemical analysis was no more than idiosyncratic observation, then there could be no guarantee of amelioration and no distinction between the real water cures and the quacks.

But geographers who study the history of science have pointed out that science has actually never traveled well, arguing that the location of a scientific endeavor makes a difference to the conduct of the experiment and that location—even of laboratories—affects the content of the results. The concept of a geography of science runs counter to our intuition because it is believed that science is a universal undertaking, not a provincial practice.[7] This same intuition informed late eighteenth- and early nineteenth-century visitors to spas, who tended to base their decisions on the infallible, universal facts of health while simultaneously holding the view that the most provincial, most place-based, least transposable characteristics of a spa would equally contribute to remaking their bodies. This contradiction of modern thought generated the intellectual ferment from which were erected curative places, most notably water cures, on the face of European and American rural landscapes.

One of these curative places was the Western Health Reform Institute, which opened in 1866 after three years of nationwide rallying and planning among the Seventh-day Adventist leadership. It was the first of many iterations of a place that can be characterized as equal parts hospital, retreat, hotel, university, and homeopathic clinic. In those early years Ellen White wrote substantially for the Adventist journal *The Health Reformer* on topics ranging from the dangers of caffeinated drinks to the benefits of loose-fitting clothing.[8] Despite her direct engagement with the journal, when it came to the institute she preferred to retain the role of visionary, keeping a distance between herself and the day-to-day implementation of that vision. Ellen White needed someone to lead the medical operations of the institute, a spokesperson and arm for her vision to change the eating, hygienic, and sexual habits of the world.

When the doors of the institute opened in 1866, the man chosen for the job was Horatio S. Lay. Starting his medical career as an allopathic (mainstream) doctor in Allegan, Michigan, Lay joined the Seventh-day Adventist church in 1856 and became interested in alternative medical practices. When he heard about Ellen White's health vision in 1863, he was one of the first to encourage her by pointing out the similarities of her vision with that of many other contemporary health reformers, notably Sylvester Graham and William Alcott.[9] The relationship between Lay and White was forged deeply enough that White hired him as first director of the Western Health Reform Institute. In the institute's first four years, Lay and his small nursing staff of Adventist farm women—donating their time from the fields—treated hundreds of patients with a series of hydrotherapeutic baths and showers.[10] Coupled with a regimented vegetarian diet served only twice a day (allowing food to fully digest), White's brand of "the natural cure" had found a place where it could be consistently practiced. A circular describing the institute stated that "*no drugs whatever will be administered,* but only such means employed as NATURE can best use in her recuperative work, such as Water, Air, Light, Heat, Food, Sleep, Rest, Recreation, &c. Our tables will be furnished with a strictly healthful diet, consisting of Vegetables, Grains, and Fruits, which are found in great abundance and variety in this State."[11]

After its grand opening, lodging at the institute quickly filled to capacity. Lay and White seemed to have tapped a nerve, a widespread desire among the Adventist followers to be taught how to behave to achieve an earthly, corporeal state of being most conducive to spiritual longevity in the afterlife. Construction of the institute created a place in which a lifestyle defined by specific health principles could be easily and unabashedly attained. The institute brought together "proper" building design with its own experts in hydrotherapy and diet, and surrounded its patients with other like-minded people seeking to live systematically and healthfully. White and Lay intentionally severed the institute's practice of being healthy from medical science. They used methods ranging from outdoor gardening to dress reform instead of surgery and medicinal drugs. They taught people to avoid processed sugar and alcohol, instead serving fruits and grains at every meal.[12]

Ellen White's impulse to turn a two-story Battle Creek farmhouse into a health retreat-clinic was not without precedent. Hers was a particularly geographical solution to the problem of how one might practice being healthy. Her answer, like that of her predecessors, was that the practice of being healthy benefited from having the necessary technologies and knowledge bundled together in the same place, a place that was necessarily an alternative from high-society Americans' way of living. Daily routines for patients at the Western Health Reform Institute—and later the Battle Creek Sanitarium—had to diverge significantly enough from what they were used to such that they felt divorced from their own culture, and embedded temporarily into a new one. For example, upon his departure from the sanitarium in 1890, one visitor reported that "electricity is extensively called into use. The Swedish movement is also applied by ingeniously constructed machinery and well trained attendants. Massage, pneumatic and vacuum treatment, and sun-baths are resorted to. No treatment, in fact, is discarded which can be successfully utilized."[13]

This pronouncement describes a series of practices about which other potential visitors might have heard rumor, but not to which they would have had regular access. Electric shock, vacuums, and sun baths were newfangled technologies that drew attention to the sanitarium as a place where one could experiment with these healing machines in the care of a professional staff.

HYDROTHERAPY INSTITUTIONS

The most immediate and inspiring precedent for White came from the establishment where her first lead physician, Horatio Lay, received his hydrotherapy training. When Lay departed from allopathic medicine and sought out alternative practices, he went to Dansville, New York, to a water cure known as "Our Home on the Hillside."[14] Opened in 1859 by James C. Jackson, the founding member of the American Vegetarian Society, Our Home employed exercise, fresh air, temperance, healthy foods, and pure water to treat patients. The essence of Jackson's water cure is found in his outright rejection of the substances and preparations that were commonly used in the mid-nineteenth-century practice of medicine, a position attractive to Lay and White. Born in an era when chemicals tended to harm as much as cure patients, Jackson's etiology was founded on balancing the body's "vital" forces.[15] In other words, diseases were not substances inside the body that awaited expulsion at the hand of a physician with drugs and tools, but rather were the manifestation of an overall "morbid condition" brought on by poor lifestyle choices, like the consumption of caffeine or alcohol, too much sex, or wearing clothes that fit too tightly.[16] As Jackson himself hyperbolically wrote: "In my entire practice I have never given a dose of medicine; not so much as I should have administered had I taken a homeopathic pellet of the seven-millionth dilution, and dissolving it in Lake Superior, given my patients of its waters."[17]

Instead of medicine, Jackson used a system of water treatments to cure his patients, including cold plunges, douches, swimming exercises, and linen wraps. His water cure was one of hundreds of similarly conceived establishments that had been sprouting up throughout New England since the 1840s, each a variation on a theme that found its roots in the nineteenth-century renaissance of ancient water therapies.[18]

In the modern history of hydrotherapy, this renaissance is credited to the Austrian Vincent Priessnitz, who systematized and popularized the use of water to treat an astonishing range of ailments. He was initially convinced of the healing powers of water after he cured himself of a number health problems he endured as a young man working on a farm, including a sprained wrist, rash, fever, crushed thumb, and broken ribs. He found that applying the right combination of running water, sponges, sprays, and submersion baths consistently healed himself, others, and even, on occasion, cattle.

After gaining some local attention and building up a small following, he moved his practice to the heart of the Sudetes mountain range, to what is now known as the town of Jesenik, Czech Republic, 150 miles east of Prague.[19] It was in the Sudetes—the "Golden Mountains" near the border of present-day Poland—that Priessnitz realized his patients fared better when they came and stayed with him as opposed to simply receiving the intermittent house calls that he had grown accustomed to making. When he could gather his patients and watch them consistently, controlling their diet, movements, and distractions, there was less a chance that his water cure would fail. And so, in 1826, surrounded by freshwater springs and a gigantic forest of pines constituting a "natural inhalatorium," Vincent Priessnitz opened the first modern hydrotherapeutic institution.[20] Fifteen years later the institute had grown to accommodate over 1,600 patients, including an "elegant group" of Eastern European princes and princesses, barons, an archduchess, and military men of all ranks.[21]

The general procedure of operations at Priessnitz's institute bears a striking resemblance to what would become the Battle Creek Sanitarium, worlds and decades away in southern Michigan.[22] Its location in a bucolic landscape, summer camp–like schedule of activities, large-group dining, presence of an influential leader, and displacement of daily habits with new, healthier ones characterized not only Priessnitz's water cure but the institutions like it that would appear in Europe, the United States, and throughout much of the Western Hemisphere over the remainder of the nineteenth century.[23] Priessnitz's water cure found footing in the United States, where water's powers to cleanse, purify, extract putrid matter, soothe, cool, relax, and stimulate were increasingly promoted and sold as more beneficial than mainstream medicine.[24] Though few in number compared with European patients, the elite American visitors who went to Jesenik opened a channel of communication between Priessnitz and like-minded American practitioners in the eastern states. From this lineage of imitators emerged James C. Jackson in Dansville, the touch point for Ellen White and the Adventists.

In 1864, about a year after Ellen White's foundational health vision, she and her family made their first trip to the Jackson water cure in New York, where they stayed for three weeks, absorbing everything they could from Jackson's treatments and lectures. When White experienced Jackson's rural water cure establishment, she recognized it as a model that could be used to reify her health visions. From the perspective that history offers, it is not terribly difficult to observe that the contents of the visions, and the impulse to perform them in making the Western Health Reform Institute, were the products of her being embedded in a greater cultural milieu of health reform.[25] Though Ellen White consistently claimed that she had only read the likes of Sylvester Graham, William Alcott, and James Jackson after having independently received holy visions from God with the same message, it is impossible to believe the pure coincidence.[26] Instead, White and the Adventists found that an obsession with alternative health practices was the perfect ally for an alternative religious sect trying to make itself a relevant cultural force. The struggle to promote a new brand of health mirrored the struggle to establish a new brand of Protestantism. Though not stated explicitly by Adventist leaders, who at least outwardly remained driven by spiritual aims, the politics of growing a membership was served by yoking these challenges together. In a move that separated them from adherents to other new American religions of the same era (Mormons, Christian Scientists, and Jehovah's Witnesses), the Adventists were able to define themselves on two fronts—religious alterity and health alterity—thereby casting a wider net and gaining a greater hold than they might otherwise have: if you believed the Adventist health message, you had to believe the Adventist spiritual message, and vice versa. As one historian of science put it, "the spread of hydropathy became another example of the readiness of Americans to accept anything new," suggesting that experimentation with health doctrine and religion went hand in hand.[27]

JOHN HARVEY KELLOGG'S GREAT TRANSITION

Though the Western Health Reform Institute enjoyed initial success as a community of health and spiritual practice, the number of patients began to fall after its first few years, and by 1869 the institute was on the verge of going out of business. After much debate within the leadership of the church about what should be done to keep the institute solvent, White decided to push forward with fundraising campaigns among the various Adventist congregations. Once again citing divine vision, she and her cohort—which included her husband, James, and a handful of other men—were able to raise enough money to stave off closure. One of these other men close to White was John Preston Kellogg. Converts to Adventism in 1852, shortly after the sect had initially rooted itself in Michigan, John Preston and his wife, Ann, had found a spiritual home with the Adventists following their westward migration to Tyrone, Michigan, where they had become farmers. The Seventh-day Adventist message attracted John Preston and Ann, who kept a strictly religious household but were in search of a congregation to which they could devote their spiritual energies. Almost immediately upon their conversion to the Adventist faith, John Preston donated a significant share required to move the denomination's printing press from Rochester,

New York, to Battle Creek, Michigan. With the press serving as the most important tool of the Adventist's evangelical operations, John Preston cemented his good relationship with church leaders by funding its relocation. In the early 1850s, along with hundreds of other Adventist families from around the country, the Kelloggs relocated to Battle Creek, and by 1855 the small Michigan town had become the headquarters of the sect.[28]

With the support of John Preston and other Adventist backers, the financial future of the Western Health Reform Institute, at least for the time being, remained secure. One of the conditions for moving ahead with the institute was that its director, Horatio Lay, would be fired. His leadership had proven ineffective, and his grasp of holistic medicine, vision, and ability to work long hours came under critique. Though Lay had developed close bonds with White prior to his tenure as lead physician, he was a relatively late convert to Adventism, and was considered, at least by White, "too proud and self-centered for his position."[29] In 1870 the church hired Dr. J.H. Ginley to replace Horatio Lay as head physician of the Western Health Reform Institute, after him a man named Harmon Lindsay, and after him William Russell. This revolving door of leadership contributed to but also reflected conditions at the institute, which by 1875 was treating only twenty patients. Along with the problem of uneven funding, diverging views between the church leadership and the medical staff left the institute without a clear vision. In 1872, as the signs of declension were becoming obvious, Ellen White's husband, James White, articulated that the institute "could not succeed unless it was staffed by the most thoroughly trained physicians," a position that would prove wise in sustaining the institute.[30] To communicate with the masses they would need to speak the more common language of science, if only to use it for proving the veracity of Ellen White's divine visions.

Between 1870 and 1876, as the institute hung on by threads, Ellen White and the Adventist community continued to refine and promote the doctrines representative of Seventh-day Adventism, mostly relying on the journals *Review and Herald* and *The Health Reformer* to state their positions. They were building a religious and cultural identity, and setting the conditions for their next big move. During these six years, despite the uncertainty of its future, the Western Health Reform Institute became the primary icon of the sect, the outward-facing symbol of its mission to save society.

Borrowing architectural designs from the water-cure craze and combining them with the philosophy of temperance found in the health reform movement, Adventists possessed something with cultural traction. Advancing a health practice was the church's greatest opportunity to continue growing. If they were not saving the world quickly enough with their publications, then it was the Western Health Reform Institute that "would be the means of bringing thousands to a knowledge of present truth."[31]

For Michigan Adventists in the early 1870s, "bringing" thousands to the truth quickly became a nonfigurative mission. When the Western Health Reform Institute was chosen as the gem of the sect—the initiative to which a majority of revenues and energies would be channeled—leaders began to strategize on how to bring potential converts to their own confines, where they could proselytize their own message, on their own terms, in their own place of healing. The Whites recognized that for this to happen, the patient capacity of the building would have to increase, along with the number of staff and medical personnel. After the group of leaders that had to this point directed the institute, White also understood that the next director would have to come from within Seventh-day Adventism, a homegrown doctor who was already steeped in the church's health message and spiritual practice. The next leader would need to make real the vision that White and the Adventists had for the institute, creating a national destination that would bring people to a place of healing where they would be trained in Adventist health principles, and would find it hard to refute the spiritual advantages being espoused on site. As the financial stability of the institute began again to flounder, they needed not only a qualified new medical director, but a charismatic personality who could sell health and sell the religion.

Defining the Adventist identity as separate from that of other health reformers during a time of health reform crazes required a special flair, something that would set it apart, that would differentiate its product from all the others. As the Western Health Reform Institute approached its tenth birthday, the Whites recognized that a bold business move was in order. Their move was to hire John Harvey Kellogg, the freshly minted Adventist doctor, to take over as director of the institute. John Harvey Kellogg was John Preston and Ann's youngest son, who in 1876 agreed to take the reins of the institute as its next and, as fate would have it, final director. Because of the relationship that John Preston had built with the Whites, Ellen came to know John Harvey when he was young, and saw in him the potential to carry the legacy of her vision of health, while James saw in him the potential to serve as a great leader.

The Whites supported Kellogg's education as a doctor, sending him, along with three other young, male congregants, to Florence Heights, New Jersey, where this new crop of Adventist doctors attended a hydrotherapy school operated by Russell Trall, a friend of Adventism and a regular contributor to *The Health Reformer.* Trall's "Hygieo-Therapeutic College" was a six-month course that focused exclusively on hydrotherapy, diet reform, the "vital action" of life (as opposed to organic chemistry), and a rejection of pharmaceutical cures. What he taught at the college closely paralleled the matrix of alternative health beliefs that the Adventists had pieced together from Graham and Jackson, and therefore seemed a safe, trustworthy educational enterprise as they moved forward with their plans to remake the Western Health Reform Institute.

But when Kellogg arrived in 1872, he recognized that the college would not satisfy his standards for his own education in medicine. It reeked of quackery, with no formal degree offered, an inconsistent curriculum of courses, and a faculty of only three instructors. Showing at a young age the confidence he would exhibit his entire life, he was not satisfied with the paradigm of holistic, hydrotherapeutic medicine that he learned in Florence Heights. Surprisingly, this dissatisfaction was shared by the Whites. Despite their misgivings about medical science, they predicted that a solid foundation in science-based medicine would be important if they were to communicate their alternative message to a wide, popular American audience. They knew they had to be in tune with mainstream medical discourse if they were to convince people of its weaknesses. In this reaction we see some of the hedging so typical of late nineteenth-century medical beliefs. With the Whites driven by their plan to resurrect the Western Health Reform Institute, Kellogg driven by the desire for intellectual stimulation, and everyone driven by the church's spiritual mission, John Harvey continued his education at the University of Michigan's medical school in 1873–1874. There he took classes in chemistry and anatomy, and was trained in the art of surgery by former Civil War medics. After a year at Michigan, Kellogg finalized his formal training in medicine by spending another year at the

Bellevue Hospital Medical College in New York City, where faculty were among the earliest to adopt bacteriology, antiseptic practices, and instruments such as the stethoscope and microscope.[32]

John Harvey Kellogg took over directorship of the failing Western Health Reform Institute in 1876. In the span of one year he cleared the debt accumulated by the institute and started to turn significant profits. Kellogg's success can be attributed to his vision to overhaul the institute, "to turn the poorly equipped Battle Creek water cure into a scientifically respectable institution where a wide variety of medical and surgical techniques would be used."[33] As the product of their leadership vision, the Whites gave Kellogg—at least for the time being—full support to do what he saw fit.

There was no one in the Adventist community, and arguably no one in the United States, more in tune with the underlying anxieties about health in contemporary society as was John Kellogg in the last quarter of the nineteenth century. Steeped in both the conservative, Christian philosophies born from the health reform movement as well as a practice of medicine that saw the body as a collection of parts under constant attack by bacteria, Kellogg walked the line between traditionalist and modernist, neither fully subscribing to, nor ever letting go of, either one. While he claimed that "the sanitarium is a product of modern medical progress, and ought to represent rational medicine in its most advanced and most progressive form," on the next page he stated that health comes "as the result of the working within the body of that occult force recognized by the ancients, but apparently almost forgotten in modern times, the *vis medicatrix Naturae*."[34] A synthesis of these two philosophies was acted out at the Battle Creek Sanitarium in the most dramatic way by the patients and their lead doctor. Kellogg used the language and method of scientific medicine to inform, but also to explain, the "natural" eating procedures for which he would become best known.

The drama of reconciling these two philosophies came at a time when medical associations and accrediting agencies were still being established, and when going to a hospital still spelled the gravest likelihood of death. Kellogg arrived, that is, precisely when the American public needed him most. Rationalizing nature as a healer for the skeptics while demystifying the technics of modern medicine for the uneducated—and all in a safe campus setting where impending death was not a real threat—he was able to assuage everyone's concerns at once in an entertaining and educational way. Proselytizing Adventism was about to become an enterprise of selling health, selling an ideal body that would be made by eating just the right foods in just the right place in just the right way. This institutionalization of body crafting is what Foucault called biopower. The power, in this case, eventually became linked with profit motive as cereal would sell a promise of healthy intestines.

THERAPEUTIC PLACES AND LANDSCAPES

Inventing his own path through the liminal zone where tradition meets modernity, Kellogg succeeded in taking something familiar—eating—and combining it with something mysterious, rhetorically powerful, and seemingly foolproof—science. While this move is predictable to those familiar with American and European histories of science and society in the nineteenth century, what is entirely unexpected about this move is the role of regional geography in understanding how eating—and digesting—worked in Battle Creek. The infrastructural and economic geographies of rural southern Michigan would play key roles in Kellogg's path through the paradox of environmental health versus scientific health, a path that not only would revolutionize eating on site in Battle Creek, but would change the national breakfast throughout the following century and beyond.

Nineteenth-century American health reform had to happen somewhere—the phenomenon could not have existed only in the textbooks of its leaders. As is shown throughout this book, the experience of participating in the alternative lifestyle that the institute—and later the Battle Creek Sanitarium—offered, could not be contained by the walls of the buildings. This is not to say that elements of the lifestyle promoted at the sanitarium did or did not spread, but rather that the experience of going through daily schedules of baths, exercises, and regimented meals depended more on local and regional environments than one might initially suspect. While on the one hand it functioned as a safe house, an escape from the ills of a fearless, secular society, on the other hand the Battle Creek Sanitarium and the bodies it produced could not separate themselves from the landscapes and the technologies with which they were embroiled. Battle Creek and its hinterlands—the urban and rural landscapes themselves—were as much a part of the revolution in eating and digestion as were Ellen White's theology and John Kellogg's subsequent philosophy of health.

The environmental setting in which patients found themselves when they deboarded the train in Battle Creek is central to understanding the tradition of retreat healing that had grown in Europe and the eastern United States since Priessnitz. Marking themselves decidedly apart from any urban connotation, health institutes manifested for their guests an idea of curative nature, an orderly, rhythmic world that should be replicated in one's daily activities. The landscape in which the sanitarium was located imparted health to its patients. When Kellogg outlined the topographic and climatic conditions for an "ideal" sanitarium (and subsequently attained them in Battle Creek) he was manifesting a philosophy of health in which the environment was a force that directly, materially affected the body. "The material conditions of a sanitarium must be of the most healthful character. It must be well located. Its site must be such as will secure in its vicinity a dry and well-drained soil, and good air, free from smoke, dust, malaria, and poisonous emanations of every sort. The climate must be adapted to the class of patients to which the institution is specially devoted . . . a cool, dry climate is especially desired."[35]

The urban condition prevented any such attempt, and thus opened the need for the design of therapeutic landscapes. Priessnitz popularized and reinvigorated the ancient idea of a system of water cures, yet his true moment of inspiration was the grounding of his operation in a single place, high in the Sudates, a destination that became imbued with special meaning. Likewise, while Kellogg was modernizing health through the stomach, he was also reacting potently against the modernist strain of urban industrialization through his choice of landscape design. Getting the body just right through science needed nature that was also just right. In a seemingly contradictory way, practicing the control of nature—through medical science and manipulation of bodies—could not happen outside of a space in which nature was exalted as curative itself.[36] The sanitarium occupied "a hill top in the far end of the town commanding a view of beautiful rolling country, green fields, and shady groves almost as far as the eye can reach. The main building is a handsome structure five stories in height, mansard roof, bay windows and broad verandas, giving it an air of beauty and comfort."[37] In this description by an editor of a Chicago newspaper, the essence of the health resort genre is captured. Amid a bucolic setting, itself curated and trimmed by agriculturalists, was a stylish and comfortable respite. Buried in fresh air, soil, and vegetation, and with few other people around, the sanitarium building offered a way for its guests to escape to nature while at the same time escaping to an alternative lifestyle. It was the perfect blending of rationality and the tools of science with romantic viewscapes. Each of these factors was considered a healing agent: the practice of medicine and the power of nature.

The roots of this perspective within the culture of water cures goes back to Priessnitz. English travel writer R. T. Claridge noted, for example, that "Graefenberg [Jesenik] is in an isolated position, out of the regular track of English migrations, leading to no place of consequence. To go thither, the English must diverge from all the leading routes."[38] On the leading routes, a Londoner would not escape British society, but would remain firmly in its diaspora. Claridge seems to egg on his readers, suggesting that the point of touring is to feel like a foreigner, that the real adventure cannot begin until one leaves the "regular track," arriving at "no place of consequence." Given the context of Claridge's statement, though, the implication is that on tour one can discover something truly fantastic that is not yet known to the traveling elite. Visual beauty coupled with the ability to de-routinize, generating feelings of awe and mild confusion, characterized the locations in which water cures thrived, and the locations that attracted tourists in droves. Priessnitz removed his establishment far from the cultural centers of Prague and Vienna, and far from the standard tourist lines that emanated from those cities into the surrounding landscapes. The idea was that the effort put forth by health seekers to arrive at his water cure was paid for by the ease with which they were able to shed the layers of habits, customs, and assumptions of their home culture. The lodge—a place of healing in its own right—was nestled in a therapeutic landscape, defined as a type of landscape in which "nature is actively enrolled in relation to bodies to create therapeutic effects in the retreat setting."[39]

Critics of Priessnitz's program of health notwithstanding, it was this geographical situation—in nature, slightly disorienting, yet intriguing, educational, and recuperative—that helped his practice grow in numbers, and that set the pattern that would eventually be adopted by the Seventh-day Adventists.[40]

THE SANITARIUM AS PLACE

To arrive at that regional geography and the role it played in making bodies between two great epistemes, let us return to the years surrounding Kellogg's initiation into the leadership role for which he had been groomed. One of Kellogg's first moves, in 1877, was to rebrand the Western Health Reform Institute into something more popular, a management technique that would prove consistently effective for him throughout the years. Rechristening the institute the Battle Creek Sanitarium, Kellogg was playing with *sanatorium,* a word that by the 1840s denoted establishments known for their favorable climatic conditions and desirable topographical situated-ness for consumptives undergoing open-air treatment. Though *sanitarium,* by contrast, was not exactly a neologism (as early as 1851 the term referred to a place one sought "superior medical advice"), Kellogg's co-optation and application of the term to his operation in Battle Creek served to set it apart from the sickly reminders of pavilion halls filled with tuberculosis patients, while at the same time conjuring the sense that it was a place where one went to refine the practice of healthy living.[41]

Armed with three different perspectives of medical practice—Trall's College, the University of Michigan, and the Bellevue Hospital—the sanitarium was the architectural means by which Kellogg would augment the Adventist health philosophy that he inherited from the Whites. The *place itself* allowed Kellogg to synthesize currents from all these perspectives into his own unique brand of medicine by bringing patient lodging, outdoor lawns and gardens, a research laboratory, an experimental kitchen, lecture and dining halls, and spa rooms with extensive hydrotherapeutic technologies all into spatial proximity.[42] By bundling these elements into what Kellogg nicknamed the "university of health," he created a destination campus in the geographical spirit of Priessnitz's water cure, but with the intellectual spirit of a sputtering modernism, one that would progress with fits toward a technological utopia, then regress in panic to the safety of religious dogma. As a place, the sanitarium created what health geographers call a "therapeutic assemblage," bringing together the material, metaphoric, and inhabited dimensions of patient experience.[43] While it was a metaphor for a culturally specific way of living, the sanitarium was more than an idea—it was built into the landscape of Battle Creek, affecting the local economy and environment. When people came to the sanitarium they did not just embody a building in a remote Michigan town; rather, their bodies inhabited—became part of—the local infrastructure, and they themselves became the crucial component in making a health practice come to life: this is the Battle Creek version of a therapeutic assemblage. If the defining characteristic of the concept *place* is that it weaves together objects, ideas, and social relations in ways that never exactly repeat, then the sanitarium was a unique place that forged a new health paradigm.[44] For Kellogg, "given a proper location, the construction of our building is of first consequence. A building is wholly unfit for a sanitarium unless it has been constructed with special reference to the purpose of such an institution."[45] This meant that the entire material manifestation of medicine's late nineteenth-century ideological confluence (would it be careful stomach measurements and chemical analyses, or water douches and jumping jacks?) had to spatially coincide at the sanitarium. This included—and this is only a short list—hollow walls, large rooms, sun roofs, hardwood floors, wide balconies with landscape views, carefully planned plumbing, proper ventilation, freshwater supply, an electrical machinery department, gymnasiums, Turkish baths, arboretums, verandas, a kitchen, and a game room. Representing an architect's version of a therapeutic assemblage, the ability to exist among these attributes—patients stepping from one episteme into another as they moved among rooms throughout the day—made the sanitarium the popular destination for healing *sine qua non.*

HOW KELLOGG BROKE FROM THE WATER-CURE ESTABLISHMENT

Convinced of the sanitarium's importance for the success of the religious mission, the Whites along with numerous church investors had long hoped to attract a greater public to the sanitarium. By the time Kellogg began as director and head physician in 1876, this hope had become an immediate financial necessity; filling the rooms to capacity was a matter of survival. This is why, despite their commitment to natural cures that eschewed the use of drugs, they were willing to support Kellogg's forays into the world of "rational" medicine, where, though he did not rampantly prescribe pills, he would mingle with bacteriologists, organic chemists, and other doctors who were convinced that nature alone was often not enough to cure. By this time, however, it was becoming clear that Kellogg was much more than a pawn in the scheme of the church. He was an energetic, determined, and well-funded young adult with a plan for how to make his Adventists benefactors proud while breaking with the water-cure philosophy that he inherited from them. Three characteristics distinguished this move that brought the health operation into conversation with a wider public.[46]

First, Kellogg was in conversation with the professional medical establishment, seeing himself as a product of the establishment's medical schools, and as someone who could contribute to its research advances. Up to then Adventists had cultivated an identity regarding health that was defined by a reaction against this very establishment. Temperance from alcohol, tobacco, meats, and spices, regular exercise, pure water, outdoor work, an abundance of sunlight, and God's will were the only curative agents worth considering for Ellen White, with microbial etiologies nothing more than another way of expressing these beliefs.[47] In other words, the Whites saw bacteriology as a handy, new, popular fad, a language that, if used with skill, could overcome popular doubt that the Adventist health message had been right all along. Explaining a recovery from typhoid fever in terms of microbes, for example, was fine so long as the curative forces were still God and nature, not the removal of the harmful bacteria from the patient's body through antibiotics. Rhetorically, Kellogg was a master at using bacteriology in whatever way tended to fit the situation, continuing to highlight old Adventist maxims like the importance of household sanitation, while offering startling, new, invisible reasons for why.

In his first major publication upon undertaking directorship of the Western Health Reform Institute, Kellogg articulated his stance on hydrotherapy. Concerned that people would continue assuming the institute was another ordinary water cure, he sought to halt this conception, one that threatened to pigeonhole and sink the operation. Kellogg took direct aim at Priessnitz, stating that "Priessnitz himself was an ignorant peasant.

He was innocent of either anatomical or surgical knowledge. His slight acquaintance with physiology was gathered by cursory observations of patients. . . . His lamentable want of knowledge allowed him to fall into many errors."[48]

Here we see Kellogg from the outset aggressively trying to distinguish his newly acquired institute from the legacy of hydrotherapy. He recognized that the architecture of healing—the water-cure spa—that White inherited from Priessnitz, Trall, and Jackson was due for an overhaul. By 1876 Kellogg already possessed many of the skills he would later use to make such an overhaul—for example, training in bacteriology—but he did not yet have a clear sense of how his break from hydrotherapy would be practiced on a day-to-day basis. That is why the remainder of his 1876 book *The Uses of Water in Health and Disease* retreats to a Priessnitz-inflected instruction book on hydrotherapy. A full ninety pages, for example, is devoted to the "applications of water," resembling the advice given by his water-cure ancestors. This served to fluff up his curriculum with borrowed material, one for which the real contents were still being conceived. When Kellogg did yoke his own philosophy to the bodies of his patients, the results were inevitably experimental. For example, the increasing availability of electricity in the 1880s prompted Kellogg to explore how it could be used in the creation of new health technologies, making possible the introduction of machines such as the Light Bath, Electric Bed, Mechanical Gym, and Vibro-Therapy Bed (see figure 1). Objects that have inspired derision since their release, they nonetheless were brilliant in capturing the progressive spirit that sought to apply technological advances in all imaginable spheres—in this case, systematic, holistic health.[49] Exemplifying a design solution to Kellogg's synthesis of traditional and modern, the electric machines marked the beginning of a new wave of alternative health, one defined by machines increasingly distant from the site of healing, the body.

Kellogg's second major break from water-cure culture involved bringing one of its distinctive traits—food and diet—to the forefront of the patient experience. It was customary for operators of hydrotherapy institutions to make a statement about the food that guests would be served during their stay. Approaches to eating varied among, and even within, institutions, ranging from attractive banquets meant to lure guests into a long stay, to unseasoned grains as part of a structured abnegation of desire

Figure 1. Electric beds, such as this, were meant to passively exercise immobile patients using electric currents. Photo taken by the author at the Seventh-day Adventist Heritage Center, Battle Creek, Mich.

for worldly excess. While the diet one encountered at a water-cure institute reflected the philosophy of its director, it was seldom granted the healing properties of water. The English traveler R. T. Claridge reports that eating at Priessnitz's institute was a gluttonous affair: "the dishes generally served at Priessnitz's table are meat, soup, boulli, with horseradish . . . veal, mutton, pork, venison, ducks, and fowls with plum sauce and potatoes; all kinds of pastry, and some vegetables, but in less quantity than meats. . . . He gives them solid, coarse, and indigestible food, in order to inspire them with courage and confidence . . . these invalids can digest those things which, in health, they would not venture to eat."[50]

In Jesenik, this rich, adventurous eating went along with conspicuous eating. The sumptuous feasts made Priessnitz's institute a society destination, where aristocrats and military men could perform their knowledge of foodstuffs in a setting that promised to refine their health in a new way.

Nearly thirty years later at the Our Home water cure in Dansville, New York, on the other hand, James Jackson took a more pious approach to food, stating that "gluttony is a very great cause of disease [that] depends largely upon the kinds of food eaten." Jackson's great move away from Priessnitz was to conceptually link diet with disease: "if food and these other agents [i.e., water] were properly used, persons would not have these diseases."[51] The diseases he lists include nearly every health issue with a name, from blindness to heart inflammation to paralysis. For Jackson, who was Ellen White's immediate predecessor and inspiration for the Western Health Reform Institute, the regulation of diet was a general panacea for physical and spiritual deficiencies of any type. The difference in Jackson's approach to food stems from his affiliation with the Vegetarian Society, an ally of the health reform movement. These groups represented a particularly American point of view on food, whereby eating was a device that brought spiritual practice into the mundane fold of sustenance. Food was an important consideration when organizing any water cure, but—despite Jackson—it nonetheless remained more of a requisite for satisfying appetites, not an active participant in health; water cures were, after all, water cures, not food cures.

This approach to food would be revolutionized by John Kellogg at the Battle Creek Sanitarium during the last two decades of the nineteenth century. When it came to food, Kellogg saw vast, uncharted fields of research surrounding diet and health that had only been hinted at by the boarding directors of other place-based health resorts. His contribution was to transform food into medicine, an act that would definitively influence the architecture of healing at the sanitarium, adding experimental kitchens and chemical laboratories to the collection of other, more traditional water-cure amenities. The pharmacological use of food replaced the drugs that were forbidden by the ethos of the health reform movement. This position remains strong in the American discourse surrounding health even today, with countless examples of the moral and corporeal superiority of "natural" antidotes over chemically engineered medicine.[52] For Kellogg, foodstuffs became tools of precision that would act on specific ailments for individuals under his charge. He tailored eating instructions for patients with severe enough health problems, going so far as to assemble a kitchen team that would innovate custom prescriptions on a case-by-case basis. Far beyond an invocation of food's vague correlation with health, Kellogg's approach was the leading edge of what, over the course of the twentieth century, would evolve into a scientific knowledge of food nutrition, and a cultural obsession with eating right.[53] Turning a water cure into a food cure meant that Kellogg would systematize food production and consumption such that bodies were remade with the intention of the physician, leaving nothing to chance on any plate at any meal.

Intertwined with his focus on eating, the last and most distinctive way that Kellogg diverged from traditional water-cure establishments was to introduce special attention to the process of digestion. Not unlike food, digestion—or, more accurately, *in*digestion—was a topic of perennial concern in nineteenth-century medical writings, with food- and water-borne disease frequently leading to chronic pain and death. The logic that Kellogg used to prescriptively assign foods as cures to maladies was the same logic he used on the human body itself. That is, Kellogg transformed the concept of indigestion from pain that was the vague effect of a poor relationship with one's environment to a precise problem existing in mapped locations throughout the innards of the body. His medical training gave him the skills to parse the constituent functionalities of patients' bodies so that the body was no longer a single thing made ill by the forces of nature, but a collection of parts that could be singled out, separated, and treated with pinpoint accuracy.

For Kellogg, the group of organs with the most potential to affect change in other parts of the body, and to affect one's overall state of being, was the digestive system. By making the digestive system an object in and of itself, the entire concept of the human body moved from a self-contained whole to a box of puzzle pieces that could be studied individually, about which separate hypotheses and conclusions could be drawn, and which could fit back into the whole in altogether new ways. As a fulcrum for leveraging healing to the entire body, the digestive system was drawn out from its body ecology and placed on its own. And as we shall see in the coming chapters, this segregation from the rest of the body meant that the digestive system was placed in an entirely new network of technologies that extended spatially throughout the hinterlands of southern Michigan. Executing the isolation of the digestive system put it in contact with the outside world.

CONCLUSION: THE SANITARIUM EXPANDS

Understanding the Battle Creek Sanitarium as a *place of healing* connects it aesthetically and experientially with its predecessors in modern hydrotherapy that began with Priessnitz and continued through, among others, Trall and Jackson. In this context it is not so strange that the sanitarium came into existence in the form that it did, nor is it strange that the "natural" methods of healing—healthy eating, calisthenics, walking, light baths, water sprays, and massages—were adopted with such enthusiasm. The precedent had been set for those who had either already attended or read about such health vacations. From published travel accounts, people were able to imagine what to expect, and how one might behave upon arrival at a water cure.[54] But while Kellogg was in part familiarly reproducing what he had already seen, he augmented the typical water cure so dramatically that a new type of institution emerged. This new type of institution would address head-on the modern contradiction that at once heralded the curative properties of nature-in-place and the development of a place-less scientific thought. As stated in a sanitarium advertisement from 1892, "if, in addition to rest and recuperation, the invalid needs a careful study of his diseased conditions, and an intelligent regulation of diet, exercise, and all other health conditions—in other words, scientific health culture . . . the Sanitarium at Battle Creek, Michigan, many years ago undertook to organize a thoroughly scientific institution which would represent rational medicine in its most advanced form."[55]

Contrast this stance with the primary message of the journal that Ellen White started in 1866—*The Health Reformer*—the very motto of which was "our physician nature: obey and live." It would be tempting to think that when Kellogg took over the sanitarium in 1876 a dramatic shift occurred from a naturalistic health philosophy to one rooted in science. But in John Kellogg's 1891 lecture "Nature's Method of Defending the Body against Disease," he says that "we may introduce many dangerous things [germs] into the body without stopping it. Why? Because the body has the power of adjusting itself to these things: it is a self regulating machine; it can constantly adjust itself to changes."[56]

Kellogg refused to give up on either of these viewpoints, blending provincial naturalism with the absolute scientific to generate a new patient experience. Priessnitz's contribution had been to systematize and categorize the ancient practice of taking waters, giving hydrotherapy a name, a new architecture, and a book with a set of instructions.[57] Kellogg's contribution was to peer deep inside the body without cutting it open, manipulating the body-machine with chemistry and food.

2 Scientific Eating

KELLOGG'S PHILOSOPHY OF THE MODERN STOMACH

THE STOMACH AT THE CENTER OF HEALTH

In 1896, during John Kellogg's prime professional years as director of the Battle Creek Sanitarium, he published a book called *The Stomach: Its Disorders and How to Cure Them.* Although he published prolifically throughout most of his career, *The Stomach* stands out as the keystone articulation of his approach to the human body, and gives the tidiest rationale for his sculpting of the cuisine that would eventually help define popular American health food. Stated most simply, the response of patients' digestive systems controlled the iterative process of designing health foods in the experimental kitchen at the sanitarium.[1] Through chemical analysis, staff quantified the response of the stomach to controlled alimentary inputs during a patient's stay, measuring, among other elements, the level of "organic hydrochloric acid" in the stomach after eating. By conceptually drawing the organs of digestion—the stomach, intestines, and colon—away from the rest of the body, health care practitioners at the sanitarium made real Kellogg's definition of the human body as "a stomach with various organs appended." Since he considered the stomach the center for the nutritive processes of the body, "any derangement of its functions must therefore result in disorder of the entire organism. . . . The great majority of diseases are primarily due to derangement of the digestive processes."[2]

Digestive organs serve as a major material interface between body and outside world, violently disintegrating plant and animal matter, transforming them into the proteins that comprise an organism. The digestive system is, in this way, the set of organs that provide the raw, material connection between human bodies and the flora and fauna that become food. Given that the incorporation of nature into human bodies cannot happen without digestion, Kellogg's act of isolating the digestive system was an explicit proclamation that treating disease relied on controlling diet, and therefore an implicit proclamation that the quality of bodies depends on the quality of the environment from which foods come. Centering a health regime on the stomach, in other words, meant centering it on food, and implicitly, therefore, it meant centering it on the environment. To be sure, the power of diet had been acknowledged by many other health reformers, but none of them managed the connection between food and specific body parts with such exactness and rigor. Kellogg answered the question of what foods cured dyspepsia (a generic term for chronic indigestion) and myriad other health problems by creating a systematic program of eating for his patients.

This chapter explores the meanings of the stomach and digestion for Kellogg and other health reformers—including his wife, Ella Eaton Kellogg, who led the Battle Creek Sanitarium's experimental kitchen. The Kelloggs were early adopters of germ theory and believed that they could trace almost any illness to putrefaction of ingested food. John Kellogg's response, then, was to connect diagnoses of ill patients to particular digestive disorders—for example, digestive action that was too fast or too slow—and then to prescribe a diet specially designed to rework the function of the digestive tract.

Whereas most health reformers made distinctions between good and bad foods that would affect one's whole constitution in either beneficial or detrimental ways, Kellogg took this impulse to higher levels of precision. In an 1892 statement, for example, Kellogg offers a layman's description of what laboratory analysis does for healing at the sanitarium: "We have long been possessed of abundance of artillery and ammunition with which to fight the hydra-headed disease, dyspepsia, which is perhaps responsible for more human ills and woes than any other malady; but our efforts have been often fruitless because we were obliged to aim our artillery without precision, like firing at an enemy concealed in a fog-bank."[3]

Given his training in gastroenterology, he saw food from the micro scale of the stomach rather than the macro scale of the body.[4] He consistently sought to create a language whereby the results of eating were measured in terms of the responsiveness of the stomach so as to avoid the "fog" of uncertainty that surrounded more blunt diagnoses and cures.

In the mid-nineteenth century, most medical practitioners were generalists—that is to say, they would prescribe and administer drugs, as well as perform surgery and midwifery depending on the patient's needs, not claiming specialty over any one body part or set of parts.[5] Though many common complaints were diagnosed under the broad heading of dyspepsia, abdominal surgery was considered very dangerous and was rare. General practitioners would often avoid surgery and prescribe elaborate drug regimens, such as the following given to a patient in 1854 suffering from a digestive disorder in Burlington, New Jersey:[6]

Day 1: castor oil, ginger, opium
Day 2: potassium citrate with lemon syrup
Day 3: opium and ipecac
Day 4: opium and ipecac
Day 5: potassium citrate, lemon syrup, nitre
Day 6: potassium citrate, tincture of aconite
Day 7: ipecac and opium
Day 8: calomel, ipecac, opium
Day 9: castor oil, ginger, orange
Day 10: gentian compound, kino pills, opium
Day 11: lead acetate, opium, bicarbonate of soda

As patients more frequently eschewed such seemingly caustic regimens, Kellogg's drug-free diet prescriptions appealed to those who sought treatment "in accordance with laws of health, approved by the scientific world."[7] Individualizing those diet prescriptions relied on careful examination of the digestive organs.

PEERING INSIDE THE STOMACH WITH CHEMISTRY

The intake procedure for patients at the sanitarium spanned several days of diagnostic tests, nurse visits, and at least one in-depth examination by Dr. Kellogg, all of which acclimatized new arrivals to the sanitarium culture. After a carriage ride from the train station, a patient would be ushered to a bath and a lodging room. Typically, a staff nurse would initiate a program that consisted of "skin friction," gentle massage of the stomach, naps, "hot applications," and sponge baths.[8] The next day she would meet with a staff doctor and nurse, who would interview the patient about her history and current complaints. The staff would then apprise Dr. Kellogg of the interview results, after which he would conduct an examination. During the exam, "every muscle, every nerve, lungs, heart, stomach, liver, and bowels" would be "thoroughly searched for diseased places" (see figure 2).[9] More often than not, the stomach was found to be the "root of all evil," prompting a "three-day study of digestion" that would determine how to proceed with the patient's treatment regimen.[10] During this multi-day study of the digestive system, Kellogg sought to translate the organs into the language of chemistry, transforming them from complex biological, material objects into a table of coefficients and quantities. After fasting the night before, the patient was ordered a test breakfast of four crackers, color reagents, and hot water. After one hour, a specially designed tube removed the contents of the patient's stomach (see figure 3), whereupon they were sent to the sanitarium's Laboratory of Hygiene.[11] Scientists in the lab measured for hydrochloric acid, "chlorin," maltose, dextrin, fatty acids, proteids, and starch.[12] These tests continued as long as necessary depending on the particular maladies and the patient's rate of healing, which could go on for months. In one case, a "senator from a western state" arrived in 1896 after suffering from indigestion for twelve years. Kellogg found a mass the size of an orange near the region of the pylorus, which he determined was not cancerous. After several months of "careful regulation of diet," the senator was claimed to be restored to good health.[13] The first and final lab results of the senator's stomach contents retrieved by the tube show an increase in acidity, maltose, and starch due to Kellogg's treatment at the sanitarium (see figure 4). These increases were considered beneficial because they helped cleanse the stomach of putrefying organic matter (i.e., food).

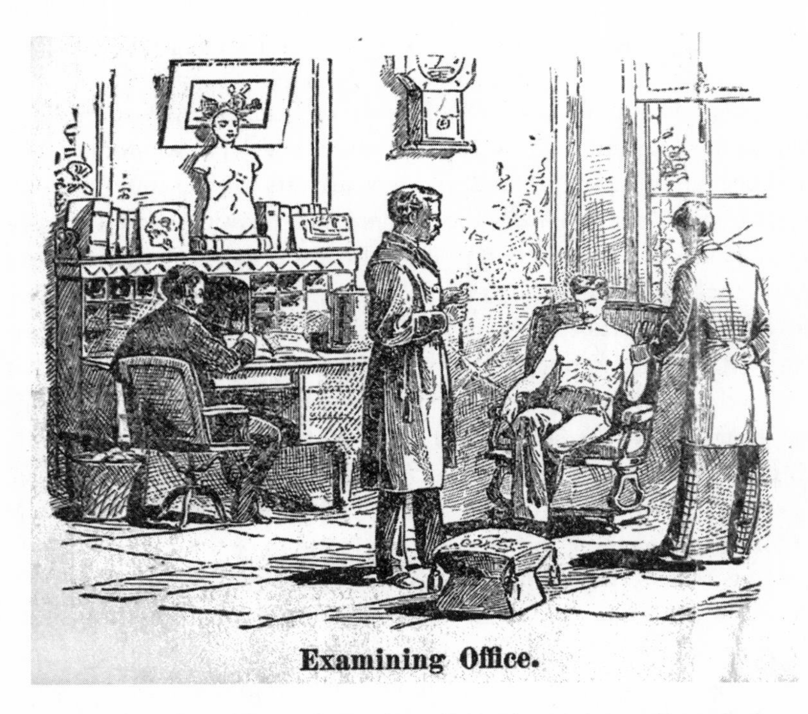

Figure 2. Portrait of the Battle Creek Sanitarium's examining office, where Kellogg and his staff would administer, among other exams, "an examination to determine the condition and position of the abdominal viscera." Image reproduced from "Good Words from the Press," ca. 1882, *Daily Inter Ocean* newspaper, courtesy of the Bentley Historical Library, University of Michigan. Digital photograph of document taken by the author.

Figure 3. The stomach tube replaced other methods of examining the contents of the stomach, including sponges, induced vomiting, and in extreme cases incising the skin to open up the abdominal area. Image reproduced from John H. Kellogg, *The Stomach: Its Disorders and How to Cure Them* (Battle Creek, Mich.: Modern Medicine Publishing Co., 1896), 321.

	No. 5018.	No. 6054.
Total acidity	.140	.164
Calculated acidity	.104	.170
Total chlorin	.296	.276
Free HCL	.000	.026
Combined chlorin	.104	.114
Fixed chlorides	.192	.106
Maltose	.272	.720
Dextrin and soluble starch	1.828	.680
Fatty acids	.055	.010
COEFFICIENTS:		
Proteids (a)	.82	.85
Starch (b)	.16	.64
Solution (y)	1.00	1.20
Motility (z)	.16	.16

Figure 4. Laboratory results obtained by sanitarium scientists upon examination of the contents of the stomach tube for a patient who had complained of indigestion during the twelve years prior to his arrival in Battle Creek. Test numbers 5018 and 6054 represent the first and last test results for this patient. Image reproduced from John H. Kellogg, "Cancer of the Stomach" (paper read before the Michigan State Medical Society, 1897), 3. Courtesy of the Bentley Historical Library, University of Michigan.

After a patient's first visit with Dr. Kellogg—following a day or two of nurse meetings, hydrotherapeutic baths, and no doubt a variety of calisthenics—and exactly one hour after she ate the test breakfast, the contents of the stomach were withdrawn and transported to the sanitarium's Laboratory of Hygiene, where chemists measured for the presence and quantities of such elements (see figure 5). The ideal stomach—the modern stomach—existed in a pristine state, undefiled by the dirty, outside world captured and swallowed with each bite of food. Ironically, then, eating at the sanitarium was, by definition, a departure from the normal course of conduct for the digestive system. Having food in the body was a temporary, imperfect state of spoliation that had to be endured, and from which a rapid recovery was engineered. Harnessing the power of chemistry, Kellogg devised a way to peer at the digestive process without submitting

Figure 5. A room in the Battle Creek Sanitarium Laboratory of Hygiene, ca. 1900. Image courtesy of the Loma Linda University Department of Archives and Special Collections, Loma Linda, Calif.

patients to the scalpel.[14] It was nonetheless an obtrusive process during which the tube was "grasped by the muscles of the throat in the act of swallowing," and was "quickly carried down into the stomach."[15] Of greater importance than a physical examination, the quantitative chemical examination of the stomach fluid gave the strongest indication for making subsequent diet prescriptions. Peering into the stomach in this way was a breakthrough—not because no one had ever tried it, but because at the sanitarium the process was routinized and applied to all patients as a matter of course, while connecting the test results with prescription diets.

There seems to have been widespread curiosity among patients to swallow a tube, siphon out the gastric juices, and discover one's state of health as indicated by laboratory scientists on a standard chart.[16] The intrigue and wonder concerning something so naturally hidden as digestion, and so distasteful in public conversation, came from Kellogg's enthusiasm. "The withdrawal of the contents of the stomach in the midst of the process of digestion, enables us to surprise the stomach at its work, and thus to detect special faults, such as excess or deficiency of activity, abnormal fermentations through the development of germs in the stomach, etc., and thus makes clear much which has heretofore been a mystery, or a matter of pure speculation in relation to the disorders of digestion."[17]

Performing over fifty chemical tests for each patient, assessing the *rate of food's movement* through the stomach drove the laboratory staff in their analyses of samples obtained from the stomach tube. Since the 1870s, popular demonstrations advocating the germ theory of disease were quick to point out that food bred bacteria outside the body. Amateurs as well as trained bacteriologists captured the public imagination by showing that the introduction of invisible organisms—"microbes" visible only with a microscope—to a plate of sweet and pure foods transformed them into a repulsive plate of slimy and putrid mush.[18]

By the 1890s, operating completely within the paradigm of bacteriology at this point, Kellogg followed the lead of a number of European scientists—including Charles Bouchard, Arbuthnot Lane, and Elie Metchnikoff—who logically, and not entirely incorrectly, surmised that food continued to breed bacteria inside the stomach after it had been swallowed. The putrid, slimy mush inside the body posed an undesirable situation that—so medical practitioners of this reasoning believed—led to self-poisoning, or *auto-intoxication.* If bacteria were vectors for cholera and typhoid, then a large number of them inside the digestive system, reproducing and accumulating exponentially, could not possibly benefit the health of the body.

Kellogg saw bacteria, or germs, as detrimental to health, and this is the crux to understanding his regime of digestion and the foods he innovated to clean out the body. As Kellogg put it, the chemical analysis determined "the number and kind of germs present, the number found varying from none in a healthy stomach to many millions in an infected stomach."[19] The insight from which Kellogg's digestive and dietary program evolved was that germs, while different from one another, were all potentially poisonous, and had to be eliminated from the body. The ways that Kellogg imagined the connection between germs and disease prepared him to adopt the paradigm of auto-intoxication. An early advocate of bacteriology, Kellogg saw germs as minute specks of life, infinitesimal in their small size; by 1882 he pronounced that despite their size, germs were "more potent for harm to human life and health than all other agencies combined. Undoubtedly these are the active agencies which give rise to the terrible typhoid fever which annually carries off thousands of victims, to dysentery, cholera, diphtheria, yellow fever, the plague, and a long list of diseases, the exact number of which is not yet known."[20]

In Kellogg's writings, and in the medical literature at large, the specific causative actions of microscopic life on the formation of disease remained murky into the 1890s.[21] Once he understood that diseases were correlated with the presence of germs, however, he rapidly began building a program whereby the stomach would be sterilized of all microscopic life. A nuclear bomb strategy, Kellogg offered his patients a natural antibiotic based on diet. He figured that bacteria came from food, and that food came into first contact with the body through the alimentary canal. What touched the stomach would be dispersed through the rest of the body.

The rate at which food was digested indicated how much time bacteria would have to multiply in the body, and therefore, in Kellogg's logic, served as a diagnostic tool for the overall health of a patient. *Stagnation* became a catchphrase in the halls of the sanitarium. All the activities, all the lectures, and all the devices revolved around minimizing the amount of time that food stayed in physical contact with the inside of the body. The root idea of stagnation can still be seen in fad diets near the end of the twentieth, and into the turn of the twenty-first, century. In their 1985 diet book *Fit For Life,* for example, Harvey and Marilyn Diamond explain the notion of "food combining," reminiscent of something that could have come directly from John Kellogg a hundred years prior. They say, "It is not only what you eat that makes the difference, but also of extreme importance is when you eat it and in what combinations. . . . Food combining is based on the discovery [*sic*] that certain combinations of food may be digested with greater ease and efficiency than others." They go on, nearly mirroring Kellogg's hygienic philosophy of eating: "the human body is not designed to digest more than one concentrated food in the stomach at the same time. . . . Any food that is not a fruit and is not a vegetable is concentrated."[22] The popularity of fad diets like this one—in which health is attained by controlling the metabolism of the gut—finds its modern American cultural roots in the practices that Kellogg organized in Battle Creek.

Kellogg's adherence to the principles of auto-intoxication came from a French pathologist named Charles Bouchard, who asked, "What are the conditions which render possible the development of the microbe? What are the conditions which may hinder its multiplication?"[23] For Bouchard, the body constantly produced toxins, major sources of which were foodstuffs and their putrefaction in the stomach. He made the tie between stagnation and self-poisoning when he observed that the intestines did in fact eliminate many toxins, but that slow bowel movement allowed for increased absorption of toxins into the bloodstream.[24]

After analyzing a patient's stomach fluid, physicians at the sanitarium classified the results into one of four main categories, signifying the amount of "work" done by the stomach in the one-hour period between ingesting the test meal and pumping out the contents. The condition *hyperpepsia* denoted excessive stomach work, *hypopepsia* a deficiency of stomach work, *apepsia* the absence of digestion entirely, and *dyspepsia* a miscellaneous category for any type of change in stomach work. Kellogg assigned subclasses to each of these four categories, ending with a total of twenty-six varieties of digestive deviance. In the laboratory, Kellogg directed chemists to measure quantities of a variety of substances, including chlorine, peptones, propeptones, albuminoids, rennet ferment, zymogen, starch, lactic acid, acetic acid, butyric acid, mucus, blood, and bile. These numbers populated a standard chart, which revealed the state of a patient's stomach in the language of chemistry (see figure 6). Food that stagnated in the digestive tract caused "fermentation." Adherents to the theory of auto-intoxication seldom claimed the possibility of a stomach devoid of bacteria, but rather claimed that fermentation—and therefore bacteria counts—could be controlled by diet. As Kellogg wrote, "when digestion is promptly performed, the food is digested and absorbed before it has time to ferment; but when it is slow, either on account of deficient muscular action on the part of the stomach, or because of deficient or defective secretion of gastric juice, fermentation takes place before digestion is completed."[25] High bacteria counts accompanied fermentation, and thus the quick movement of food through the digestive system was considered beneficial *a priori*.

NO............

SANITARIUM LABORATORY OF HYGIENE.

BATTLE CREEK, MICH.

J. H. KELLOGG, M. D., Superintendent.

ANALYSIS OF STOMACH FLUID.

M.......................... ....189..

Test-meal. Bread, fermented, unfermented....oz., meat, eggs....oz. Water, 8 oz. Breakfast. Dinner. (Lavage two hours previous.) Duration of Digestion.....hours .....minutes.

Physical characteristics: amount.......c.c., color....... odor......blood.....mucus......residue ....gms.

Normal variations.*

Total acidity, (A)....gms.(A')....gms.(.180-.200 gms.)
Coefficient, (*a*)................(.86)
Total chlorine, (T)...........gms.(.300-.340 ")
Free HCl, (H).........gms. } " (.025-.050 ") } .180-
Combined chlorine, (C).. " } (.155-.180 ") } .225.
Fixed chlorides, (F)............ " (0.109 gms.)

Color reactions for HCl: Congo-red...Methyl-violet........ Günzburg's reagent....Resorcine....Dried residue....

Reactions for fatty acids, etc.: Lactic.............Acetic or Formic .....Butyric. .....Bile......Alcohol.......

Digestion of Proteids: Biuret reaction..........purple, violet Rennet ferment..........Rennet zymogen...........

Digestion of Starch: Lugol's solution..blue, brown, violet, purple.

Diagnosis. Formula: A *a* T $\frac{H}{C}$ } ....

....

Prescription..

....

....

* The quantity of chlorine is expressed as HCl. The values given relate to 100 c.c. of stomach fluid.

Figure 6. Standard graphic representation of the health of the stomach at the Battle Creek Sanitarium. Image reproduced from John H. Kellogg, "The New Chemistry of the Stomach," *The Bacteriological World and Modern Medicine* 1, no. 13 (1892): 431. Courtesy of the Bentley Historical Library, University of Michigan.

One patient—a Mr. J. R. from Pennsylvania—stayed at the sanitarium for twenty days in 1900, during which time the number of aerobic and anaerobic bacteria in his stomach purportedly went from 100,000 and 52,000, respectively, to zero (see figure 7). It is now difficult to believe that people could live with *no* bacteria in the stomach, and as such the side-by-side results of this patient's arrival and departure bacteriological examinations point to the dubious methods of measurement in the Laboratory of Hygiene, while also pointing to Kellogg's unwavering ambition to cleanse patients' stomachs. To reduce fermentation, to give the stomach time to work on the food before it became festered in bacteria, Kellogg and his team in the sanitarium's experiment kitchen introduced "sterilized" food to the patient's diet. Using its acids, a healthy stomach could completely destroy the small number of bacteria found on grain-based foods that were "heated to a temperature sufficiently high to destroy any germs."[26] Sterilizing food, in other words, meant baking it to a crisp so that no living organism—no matter how small—could possibly survive in or on it. An astonishing variety of food products emerged from this philosophy and practice, wedding the Laboratory of Hygiene with the experiment kitchen in a way that sought to entirely exclude a portion of the outside world from the inside of people's bodies. Though this may at first seem to be the opposite of an extensible body—completely disconnected from nature—it instead shows Kellogg's hypersensitive awareness to the source, quality, and content of the foods that he served.

After chemists populated the standard chart shown in figure 7, doctors then compared it with another chart that served as a key to access the meaning of the quantifications, a foundation for the diagnostic conclusions that would determine a patient's diet prescription (see figure 8). Quantifying the action of the stomach on food allowed sanitarium physicians to codify the state of a patient's health, producing an absolute, measurable standard to which all bodies were conformed. A scientifically right stomach was the means to a perfect, reproducible, modern body, cleaned of nature and removed from nonscientific health mythologies.

The stomach's role as the biological organ responsible for incorporating food changed into a new role as a prognosticator, the organ that claimed special power in foretelling one's overall health. Kellogg's approach attempted to remove the guesswork from diagnosis, believing that lab

Laboratory of Hygiene of the Battle Creek Sanitarium

BATTLE CREEK, MICH.

J. H. KELLOGG, M. D., Superintendent. F. J. OTIS, M. D., Director.

Report of the Chemical and Bacteriological Examination of the Stomach Fluid Obtained after a Test Meal in the Case of

Mr. J. R. .... State Pa. .... Date 8/21. 9/10 1900

Test Meal. — Regular .... Time of Digestion: 1 .... 1 h 0 ... 0 m

Special .... " " .. ..h......m

PHYSICAL EXAMINATION.

Amount .... 44 185 c.c. Color .... n .... n ..

Specific Gravity 1.020 1.017 .. Odor .... n .... n ..

Residue .... 25 50 grms. (normal 20 grms.) Blood ...... 0 ... 0 ..

Estimated stomach contents 85 ... 202 c.c. (normal 120 c.c.); Mucus {Normal / Catarrhal} 0 ... 0.

MICROSCOPICAL EXAMINATION.

Blood . 0 .... 0 Misc. ....

Pus .. 0 .... 0 ....

BACTERIOLOGICAL EXAMINATION.

Bacteria, No. per c c. (30 c.c. = 1 oz.) { Aerobic .. 100,000 .... 0 ......

{ Anaerobic .. 32,000 .... 0 ....

Yeast " " .... 0 .... 0 ....

Mold " " .... 0 .... 0 ...

Acid formation ..... + ... 0 ..... Milk coagulation .... + ... 0 ....

Gas production ..... 0 ...... 0 .... Gelatin liquefaction ...... + ... 0 ...

Bread media .... Fruit Media ....

Figure 7. Results of two chemical examinations of Mr. J. R.'s stomach, upon his arrival at the Battle Creek Sanitarium on August 21, 1900, and upon his departure, on September 10, 1900. Through prescription diets, the number of bacteria per cubic centimeter was presumably reduced from thousands to zero. Image reproduced from John H. Kellogg, *What Is the Matter with the American Stomach?* (Battle Creek, Mich.: Modern Medicine Pub. Co., 1900). Courtesy of the Bentley Historical Library, University of Michigan.

results could not lie. Interpreting these lab results was subject to the organizing scheme of the "graphic method" developed at the sanitarium. How did this scheme work? The first five columns on the left are different measures of "work" done by the stomach on food. The column labeled with the letter *H* is the most telling of the measurements. It denotes the

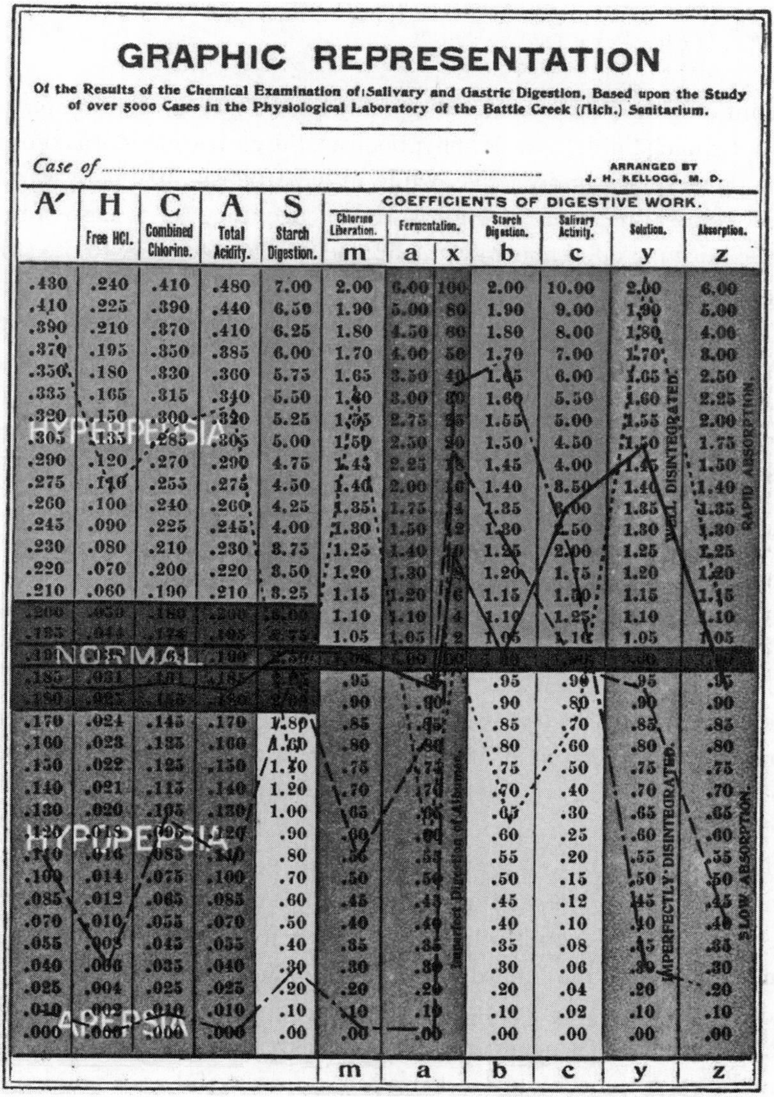

GRAPHIC REPRESENTATION

Of the Results of the Chemical Examination of Salivary and Gastric Digestion, Based upon the Study of over 5000 Cases in the Physiological Laboratory of the Battle Creek (Mich.) Sanitarium.

Case of ....

ARRANGED BY J. H. KELLOGG, M. D.

A′	H	C	A	S	COEFFICIENTS OF DIGESTIVE WORK.						
	Free HCl.	Combined Chlorine.	Total Acidity.	Starch Digestion.	Chlorine Liberation.	Fermentation.		Starch Digestion.	Salivary Activity.	Solution.	Absorption.
					m	a	x	b	c	y	z
.430	.240	.410	.480	7.00	2.00	6.00	100	2.00	10.00	2.00	6.00
.410	.225	.390	.440	6.50	1.90	5.00	80	1.90	9.00	1.90	5.00
.390	.210	.370	.410	6.25	1.80	4.50	60	1.80	8.00	1.80	4.00
.370	.195	.350	.385	6.00	1.70	4.00	50	1.70	7.00	1.70	3.00
.350	.180	.330	.360	5.75	1.65	3.50	40	1.65	6.00	1.65	2.50
.335	.165	.315	.340	5.50	1.60	3.00	30	1.60	5.50	1.60	2.25
.320	.150	.300	.320	5.25	1.55	2.75	25	1.55	5.00	1.55	2.00
.305	.135	.285	.305	5.00	1.50	2.50	20	1.50	4.50	1.50	1.75
.290	.120	.270	.290	4.75	1.45	2.25	18	1.45	4.00	1.45	1.50
.275	.110	.255	.275	4.50	1.40	2.00	16	1.40	3.50	1.40	1.40
.260	.100	.240	.260	4.25	1.35	1.75	14	1.35	3.00	1.35	1.35
.245	.090	.225	.245	4.00	1.30	1.50	12	1.30	2.50	1.30	1.30
.230	.080	.210	.230	3.75	1.25	1.40	10	1.25	2.00	1.25	1.25
.220	.070	.200	.220	3.50	1.20	1.30	[illegible]	1.20	1.75	1.20	1.20
.210	.060	.190	.210	3.25	1.15	1.20	6	1.15	1.50	1.15	1.15
.200	.050	.180	.200	3.00	1.10	1.10	4	1.10	1.25	1.10	1.10
.195	.044	.174	.195	2.75	1.05	1.05	2	1.05	1.10	1.05	1.05
.190	[illegible]	[illegible]	.190	[illegible]	[illegible]	[illegible]	[illegible]	[illegible]	[illegible]	[illegible]	[illegible]
.185	.031	[illegible]	.185	[illegible]	.95	.95		.95	.90	.95	.95
.180	.025	[illegible]	.180	[illegible]	.90	.90		.90	.80	.90	.90
.170	.024	.145	.170	1.80	.85	.85		.85	.70	.85	.85
.160	.023	.135	.160	1.60	.80	.80		.80	.60	.80	.80
.150	.022	.125	.150	1.40	.75	.75		.75	.50	.75	.75
.140	.021	.115	.140	1.20	.70	.70		.70	.40	.70	.70
.130	.020	.105	.130	1.00	.65	.65		.65	.30	.65	.65
.120	.018	.095	.120	.90	.60	.60		.60	.25	.60	.60
.110	.016	.085	.110	.80	.55	.55		.55	.20	.55	.55
.100	.014	.075	.100	.70	.50	.50		.50	.15	.50	.50
.085	.012	.065	.085	.60	.45	.45		.45	.12	.45	.45
.070	.010	.055	.070	.50	.40	.40		.40	.10	.40	.40
.055	.008	.045	.055	.40	.35	.35		.35	.08	.35	.35
.040	.006	.035	.040	.30	.30	.30		.30	.06	.30	.30
.025	.004	.025	.025	.20	.20	.20		.20	.04	.20	.20
.010	.002	.010	.010	.10	.10	.10		.10	.02	.10	.10
.000	.000	.000	.000	.00	.00	.00		.00	.00	.00	.00
					m	a		b	c	y	z

HYPERPEPSIA
NORMAL
HYPOPEPSIA
APEPSIA
WELL DISINTEGRATED.
RAPID ABSORPTION.
Imperfect Digestion of Albumen
IMPERFECTLY DISINTEGRATED.
SLOW ABSORPTION.

Figure 8. This chart represents the "graphic method" of stomach analysis employed at the Battle Creek Sanitarium. The five columns on the left give quantities of substances in the stomach as measured in the Laboratory of Hygiene. The seven columns on the right depict "coefficients of digestive work"—that is, a numerical score that compares a patient's stomach work with the "normal standard" stomach. The thin horizontal strip was considered normal, meaning that it was quite easy for sanitarium doctors to identify a patient's digestion as deviant. In the background of the chart on the left side can be read the names of the digestive diseases "hyperpepsia," "hypopepsia," and "apepsia." Image reproduced from John H. Kellogg, *The Stomach: Its Disorders and How to Cure Them* (Battle Creek, Mich.: Modern Medicine Publishing Co., 1896), 353.

amount of hydrochloric acid, colloquially known as gastric juice, present in the stomach fluid. In cases of hypopepsia, the gastric gland was thought to be dangerously underactive, while in hyperpepsia, it was overactive. Aside from eating, corrective measures to encourage the release of more gastric juice included massage, the application of electricity to the stomach, or exercise after eating. To decrease gastric juice, patients were given ice packs, galvanism to the spine, and were forbidden to consume any "stimulating" foods and condiments (e.g., vinegar, salt, pepper, meats, or sugar). This column was crucial because hydrochloric acid was the body's silver-bullet defense against "the growth of germs and the production of poisonous substances through the decomposition of the food," actions that led directly to auto-intoxication and poor health.[27]

Kellogg turned the complex digestive process of dismantling objects of nature and assimilating them to bodies into a popular, if rigid, language by devising a translational tool in the form of the standard chart. Reducing the body to its chemical measurements meant that people fit into Kellogg's program in one way or another, a convenience for the making of a business enterprise and for the construction of knowledge about the body. To put it another way: one way to imagine Kellogg's career is that it was devoted to having ready answers for any of the twenty-six possible states of digestive health in which a patient could find him- or herself. Kellogg was prolific and verbose in answering questions about the various conditions that the diagnostic chart revealed. The method that he built sought desperately to remove myth and magic from interpreting the innards of people's bodies, as he continually tried to ally himself with the medical associations founded in a rational, scientific approach.[28] Using diagnostic charts that derived their meaning from the results of chemical tests was—however useful or misguided—an objective measure, a scaffolding of information that he wielded to create his own expertise.[29] As Kellogg put it, "The ability to represent graphically by a formula the exact condition of the digestive process in a given case, and to determine mathematically the extent of deviation from the normal condition, relieves the subject of functional disorders of the stomach from the mystic vagueness with which it has been surrounded heretofore."[30]

Feminist scholars have long pointed out that the relentless search for signs of deviance is a defining obsession of the modern life sciences. In applying these signs of deviance to human bodies, signs were, and still are, visual in nature—for example, skin color and body shapes—"offering a means for controlling deviance through the clinical gaze." Medical science, in other words, gave a tool for categorizing people into self and other, normal and inferior, motivating and justifying social oppression.[31] The process of denoting deviance at the sanitarium played off this impulse, measuring digestive (mal)function so that bodies could be categorized as either healthy or ill. But there was one major difference in the medical gaze practiced by Kellogg and his sanitarium staff: the deviant body was decidedly *not* observable with the naked eye. Rather, it was cloaked under abdominal skin, shrouded in mystery. Hiding behind the objectivity of the lab result, Kellogg manufactured sick bodies by letting the chemical lab results classify people for him and his staff. When Kellogg intentionally moved away from a medical practice based in "ancient dogmas and superstitions" to "a new science,—rational medicine," the standardization process of classifying lab results as hyperpeptic, hypopeptic, and so on served as a diagnostic that spoke for itself rather than one based on the opinion of the medical practitioner.[32]

Though mostly naïvely, at least in the early days, John Kellogg and the sanitarium greatly profited from the quest for a perfect stomach as defined by the narrow strip of "normal" in the graphic representation in figure 8, thanks to shifting currents in American society. As Victorians became Progressives between the Civil War and the turn of the century, they sought answers for how to make a utopic society defined by class conformity.[33] The mission to fit oneself into the rising middle class was captured by the practice of fitting one's body into the standardizing band on the digestion chart shown in figure 8. To undergo adventures in cuisine, as well as in electricity machines and hydrotherapy, was to perform a social duty to make a society unencumbered by raging public health problems such as the spread of typhoid or cholera. Kellogg hit this cultural sentiment in stride by presenting the opportunity to practice social change in place. Arriving at the spacious, bucolic setting in Battle Creek from cities purportedly teeming with disease, most patients were already motivated to accept whatever tasks were required of them to ameliorate public health, and to separate themselves from an unhealthy population. By setting a narrow standard on the digestion chart, Kellogg capitalized on this anxiety as patients extended their stays in Battle Creek from weeks to months, refusing to re-enter society until they solved the riddle of Dr. Kellogg's program of rational digestion. While he clearly believed more than money was at stake, it nonetheless would have been impossible to grow the sanitarium into the operation that it became without a captive audience who feared urban decay, and the narrowness of the "normal" band in figure 8 into which his audience tried desperately to fit.[34]

STOMACH CHEMISTRY INFORMS FOOD INNOVATIONS

Food was dangerous, and it had to be specially modified from its natural state into a sterile, efficient one to counter the body's disposition as "a manufactory of poisons."[35] Kellogg pointed out that bacteriologists "have fully demonstrated the presence in the alimentary canal of human beings, of a very considerable number of different species of microbes which are continually producing their several poisons in greater or less quantity, the degree of their activity depending very largely upon the character of the material taken into the alimentary canal as food."[36]

An 1895 experiment conducted in the sanitarium's Laboratory of Hygiene compared stomach fluid samples from over twenty patients after alternating their test meals. The first test meal consisted of 1.5 ounces of Granose—"a thoroughly sterilized preparation of wheat"—and 8 ounces of water. In these cases, investigators failed to discover even "a single microbe," with every stomach "absolutely sterile." The next test meal consisted of browned biscuits that were not fully baked in the center, resulting in "great quantities of microbes and yeasts" in the fluid samples. Kellogg concluded from this that by using a thoroughly sterilized food, the stomach is able to destroy any lingering microbes "so that a perfectly aseptic condition of the stomach may be established."[37] From this philosophy emerged the ceaseless tinkering in the sanitarium's experiment kitchen that sought to impose nothing short of a cuisine of sterility that rested on overbaked breads in conjunction with elaborate compositions of fruits and nuts. The Kelloggs believed people became sick because they ate an infected environment. Rendering that environment sterile became the practice that emerged at the intersection of stomach biochemistry and food production.

Through the late 1880s and into the 1890s, as he embraced the writings of Bouchard, Kellogg's philosophy on eating became stricter and more exact. When the first Battle Creek Sanitarium building was erected in 1877, however, Kellogg was more approximating and forgiving to his clientele regarding matters of diet. A reporter from the early 1880s, for example, stated that "Dr. Kellogg is a rigid vegetarian, eating no meat, abjuring butter and condiments, and living on graham diet, oat-meal, milk, and fruit; but he does not insist upon placing patients on his radical platform, or more properly speaking, at the 'diet' table."[38]

It is clear that Kellogg was practicing the diet regime articulated by the Seventh-day Adventist leader Ellen White, but he was not yet mandating such a program for his patients. Coming off a period of financial distress in the early 1870s, this democratic diet table was likely a business decision to attract as many people as possible, forecasting the secularization process that the enterprise would undergo in the early twentieth century. By 1890, though, as the sanitarium grew in the national consciousness, and as patient lists grew in numbers and prestige, food became the fulcrum of health at the sanitarium. As Kellogg wrote, "Bad eating and the consequent impaired digestion, undoubtedly lay the foundation for most cases of chronic disease, and hence a careful regulation of the dietary, with the provision of suitable food for the invalid, is one of the things which certainly should be found in a well-regulated sanitarium. . . . Every patient must have his diet prescription, and must be required to follow it implicitly."[39]

The transformation of food into medicine is the hallmark of the Battle Creek Sanitarium. Local journalists of the time proclaimed that the sanitarium's diet kitchen "serves the same purpose in relation to the diet prescriptions that the pharmacy or a drug store serves in relation to medicinal prescriptions."[40] Chemistry, again, was the method by which Kellogg realigned the nature and meaning of food to fit his vision of health. Modeling his diet kitchens after chemical laboratories, he was able to quantify dosages of food items and give them to patients based on their ailments, creating health by waving the wand of nutritional science over nature. "The kitchen in such a scheme as this," says Kellogg, "must be organized as a laboratory and conducted under laboratory rules and in harmony with laboratory principles."[41] Discrete prescriptions were numbered and sent to the cooks in the kitchen, whereupon a patient's meal would be delivered to his room or the dining hall, depending on the state of his mobility. The so-called medical matron at the sanitarium noted that operations in Battle Creek offered "a specially favorable field for the study and application of medical dietetics. The absence of a regulation diet makes it possible to adapt the bill of fare to the needs of each individual patient, with a degree of accuracy which cannot be attempted under less favorable conditions. Facilities for analysis of stomach fluids and other secretions afford a basis for the exact study of the dietetic needs of patients."[42]

Each patient, that is, received her own meal, intended to correct the undesirable condition in which she found herself after being diagnosed by the medical staff (see figure 9). These meals were designed to fulfill anywhere between one-fourth and four-fourths of a patient's daily diet, the timing of their eating determined by the fragility of one's state of health.

ELLA EATON KELLOGG AND THE EXPERIMENT KITCHEN

In the confines of the sanitarium, John Kellogg sought to redesign the stomach to minimize the presence of bacteria. This design was primarily accomplished by refashioning the ecology of the digestive system through a careful series of diet prescriptions.[43] As Kellogg put it, "above all it is necessary that a careful dietetic prescription shall be made, and that the patient shall be made to carry it out."[44] In the experiment kitchen of the sanitarium, guided by the test results of the Laboratory of Hygiene, the earliest popular application of science to human food was being developed in the unlikely small, midwestern town of Battle Creek. Foundational to concocting new types of foods that answered the demands set forth by Kellogg's diagnoses was a battery of industrial appliances, including ovens, roasters, mashers, and dehydrators, machines that transformed raw foods into prepared health foods. The experiment kitchen was the place where ideas found traction, where the practice of science spilled

FRUITS, GRAINS, AND NUTS.—Continued.

	FOODS	Weight Oz.	MEASURE	CALORIES OR FOOD UNITS	
33	Granose biscuit..	6	8½	Proteids,	138
	Apples..........	12	2	Fats,	188
	Maltol..........	3		Carbo.,	956
		21		Total,	1,282
34	Granose biscuit..	5	7	Proteids,	246
	Malted nuts......	4		Fats,	238
	Peaches..........	7	2	Carbo.,	771
		16		Total,	1,255
35	Granose biscuit...	5	7	Proteid,	209
	Nuttolene..........	2		Fats,	321
	Maltol..............	2		Carbo.,	723
	Baked apples.....	6	1	Total,	1,253
		15			
36	Granose biscuit...	3	4 ½	Proteids,	221
	Nuttose............	2		Fats,	293
	Bromose...........	2	4 cakes	Carbo.,	732
	Grapes...........	18			
		25		Total,	1,246
37	Granose biscuit...	4	6	Proteids,	252
	Nuttolene..........	2		Fats,	275
	Malted nuts.......	2		Carbo.,	729
	Cherries...........	16		Total,	1,256
		24			
38	Bananas............	13	6	Proteids,	208
	Nuttose............	3		Fats,	290
	Granose biscuit...	2½	3 ½	Carbo.,	724
		18½		Total,	1,222

Figure 9. In this diet chart, prescription meals 33 through 38 are shown. Each of these meals affected the stomach and, therefore, the health of the body in a particular way. Note the abundance of Granose (baked wheat crackers), Nuttose and Nuttolene (nut-based bars and spreads), and fruits. Image reproduced from John H. Kellogg, *Balanced Bills of Fare: Arranged with Reference to the Normal Daily Ration; And the Needs of Special Classes of Invalids* (Battle Creek, MI: Good Health Pub. Co., 1899), 11. Courtesy of the Bentley Historical Library, University of Michigan.

over into a new sphere defined by the careful and numerical wedding of food to the bodies of his worried patients.

John Kellogg wrote the foundational treatise on digestion that drove medical practice at the sanitarium. He enthusiastically transferred his analytical method to foodstuffs by arranging food charts indicating the amount of proteids, fats, carbohydrates, and calories present in any single food item (see figure 10). By doing so he made the nascent field of nutritional science available to a broad audience, presaging future national, public health anxieties surrounding calorie counting, obsession with avoiding fats, conspicuous avoidance of carbohydrates, and the consumption of protein shakes and bars for body shaping. But while Kellogg articulated food in the language of science, it was his wife, Ella Eaton Kellogg, who led a team of cooks to carry out the long hours of kitchen work required to fulfill the insights and whims of the doctor. Ultimately, it was Ella who communicated most clearly the position on food adopted and promoted at the sanitarium. Preceding the publication of her husband's *The Stomach* by three years, Ella's 1893 book *Science in the Kitchen* is every bit as comprehensive about food as is *The Stomach* about digestion, and it leaves few topics untouched when it comes to answering the question "what should we eat to be healthy?" Through years of collaborating with John, fully embedded in sanitarium culture, her work as leader of operations in the kitchen allowed her to weave together John's analytical approach to food with an educational, practical take, a mix between a treatise on eating, the birth of a new health cuisine, and an instructional cookbook. Her work gives detailed, empirical answers to the question of how foods act on the body, clearly influenced by the stances developed by John and the laboratory staff. Combined, Ella's *Science in the Kitchen* and John's *The Stomach* express a revolution in American eating that irrevocably committed food to the body with strict conformity to the rules of chemistry, and offered a guide for how the American public could remake their own bodies.

After returning from an 1882 trip to Europe, John Kellogg asked Ella to lead operations of an "experiment kitchen" in an attempt to solve the problem of bland and unscientific cuisine. As its name suggests, the experiment kitchen was a laboratory-inspired place for concocting new appetizing dishes, the making of which would be carried out with the same precision as was the investigation of stomachs. Before embarking on

FOODS	Weight oz.	MEASURE	PER CENT. OF Proteids	Fats	Carbo.	Calories in one oz.
Granola	4.25	.5 pt.	15.	3.	75.	114
Granose	2.75	4	15.4	2.3	79.1	117
Zwieback	2.2	2 pieces	13.6	2.	70.	104
Graham crackers	3.	6	9.8	13.6	70.	128
W. W. Wafers	2.5	6	9.8	13.6	70.	128
Beaten biscuit	2.25	6	11.7	1.2	80.	110
Sticks	2.	12	11.7	1.2	80.	110
Rolls	4.	6	11.7	1.2	80.	110
Passover bread	3.75	12	11.7	1.2	80.	110
Graham bread	1.	1 piece	9.5	1.4	53.3	80
W. W. bread	1.	1 piece	9.8	1.4	50.7	75
W. bread	1.	1 piece	8.8	1.7	56.3	81
Nut-gravy toast	6.	1 piece	6.8	1.	35.	52
Prune toast	6.	1 piece	6.8	1.	35.	52
Berry toast	6.	1 piece	6.8	1.	35.	52
Cream toast	6.	1 piece	6.8	1.	35.	52
Nuttose roast	6.	.5 pt.	4.3	1.9	12.5	25
Nutt'se and tomato	8.75	.5 pt.	4.3	1.9	12.5	25
Potato, mashed	7.75	.5 pt.	3.2	.15	27.5	36
Vegetables, fresh	8.	.5 pt.	1.6	.3	4.	8
Tomato, stewed	8.	.5 pt.	.8	.15	1.25	3
Potato, baked	4.	1 medium	3.2	.15	27.5	36
Nut gravy	9.	.5 pt.	....	....	....	....
Soup	8.5	.5 pt.	.8	.2	1.	3
Sugar	7.	.5 pt.	....	....	97.8	111
Beans, boiled	8.75	.5 pt.	....	....	....	....
Cream	8.5	.5 pt.	2.7	26.7	2.8	75
Milk	8.5	.5 pt.	3.6	4.	4.7	21
Egg	1.5	1	14.	10.5	....	47
Ster. butter	....	... .	1.	85.	....	218
Kumyss	8.5	.5 pt.	3.7	3.6	4.7	20
Cottage cheese	....	....	3.6	3.7	6.	20
Malted nuts	....	....	23.	20.4	49.3	140
Almond cream	....	....	2.7	26.7	2.8	75
Nut butter	8.75	.5 pt.	16.4	26.4.	9.5	101
Bromose	1.	2 cakes	19 6	24.	39.4	135
Maltol	9.	.5 pt.	....	20.	75.	136
Nuttose	3.	....	30.	30.	18.	139
Nuttolene	3.	....	30.	30.	18.	139
Nut meal	3.75	.5 pt.	28.3	46.2	1.8	161
Stewed nuttolene	9.	.5 pt.	....	....	....	...
Beefsteak	....	... .	19.3	3.6	.0	37

Figure 10. A chart of foodstuffs in use at the sanitarium, including their recommended portion sizes and percent of recommended daily proteids, fats, and carbohydrates. Also included is the number of calories per food item. Today the U.S. Food and Drug Administration requires such information to appear on all food labels. Image reproduced from John H. Kellogg, *Balanced Bills of Fare: Arranged with Reference to the Normal Daily Ration; And the Needs of Special Classes of Invalids* (Battle Creek, MI: Good Health Pub. Co., 1899), 5. Courtesy of the Bentley Historical Library, University of Michigan.

this undertaking, Ella attended a number of cooking schools on the east coast. It was during this period that Ellen Richards founded the discipline of home economics, publishing *The Chemistry of Cooking and Cleaning* (1880) and *Food Materials and Their Adulterations* (1885), undoubtedly influential to Ella.[45] After nearly twenty years of directing the kitchen, Ella's greatest professional accomplishment was best articulated in the introduction to *Science in the Kitchen*. "Mrs. Kellogg has had constant oversight of the cuisine of both the Sanitarium and the Sanitarium Hospital, preparing bills of fare for the general and diet tables, and supplying constantly new methods and original recipes to meet the changing and growing demands of an institution numbering always from 500 to 700 inmates."[46]

Grounding the design of the kitchen in a way that reflected trends in home economics, Ella invented a new way of manipulating food to create a health cuisine. Good sunlight and ventilation were considered imperative, achieved by the use of windows on at least two sides of the room, designed so that they opened from the top "for a complete change of air." She emphasized this point: "The ventilation of the kitchen should be so ample as to thoroughly remove all gasses and odors, which, together with steam from boiling and other cooking processes, generally invade and render to some degree unhealthful every other portion of the house."[47] This kind of ventilation architecture used in places of healing has been cited by Foucault as a "therapeutic operator," a material agent that no longer simply provided a roof over patients' heads, but worked to change their health.[48] Ella's attention to sanitation was consistent with the rise of bacteriology. It comes as no surprise that in the field of home economics, bacteriology "figured prominently in the early discipline's teachings about a wide range of topics, including home sanitation, interior decoration, and food preparation."[49]

As John Kellogg was not one to delegate and forget, he frequently requested new ideas for foods as they came to him, and sought to involve a range of santiarium staff in the process of making new foods. To witness the relationship among the medical staff and the cooks in the experiment kitchen, one need not look further than the countless typed notes passed among John Kellogg, his brother William, Ella, the staff physicians, and the cooks. One particularly rich example is the following, written by John in 1898: "An endless number of things can be made. . . . I want to see some experiments made, also, with partially malted cereals, malting the wheat before, and then making into granose, after cooking in dry steam and drying, and also partially digesting the flour paste by adding malt meal until considerable sugar is developed, then thickening with a flour made from well browned zwieback or well baked granola, then making the cakes like granola biscuit, and grinding coarse like nuttola."[50]

Here it is easy to catch a glimpse of how they were thinking, mixing grains and nuts, cooking them and drying them into an "endless" variety of combinations. The factor giving purpose to this odd-sounding note was the food's ability to correct the work of the stomach such that a lab test would definitively show the patient's stomach to be within the limits of the standards set forth in the digestion chart. To calm, to stimulate, to sterilize, or to populate with acid, food was the scalpel of choice at the sanitarium. In the same way that he made numerous digestion charts, Kellogg also produced an entire series of food charts. By doing so he harnessed the same curiosity about the invisible, elemental properties of food as he did with the constituent chemical properties that described the stomach.

Kellogg used the diet prescriptions to act with precision based on the physician's diagnosis. Since "every cell and fiber of [the dyspeptic's] body is more or less injured by the failure of the stomach to supply proper nourishment for rebuilding the tissues," it stood to reason that giving the stomach the correct fuel would alleviate the entirety of possible ailments.[51] Diet Lists, as they were called, included courses of eating that would cure illnesses including catarrh, hyperpepsia, fever, diabetes, rheumatism, gout, and constipation, to name a few. Each diet, printed prolifically in Kellogg's 1896 book *The Stomach,* was formulated to locate precisely the cause of the patient's malady, and to act precisely on it, each one being "totally unfitted for all cases alike."[52] Designing a typology of eating that aligned food with body parts and body ailments brought an analytic impulse to health foods never before seen. When woven into the language of the patient experience, the Diet Lists gave immediate gratification, an action plan to remold the body in accordance with the laws of efficient digestion. Recipes for dishes in the Diet Lists exactly follow those made by Ella Kellogg in the experiment kitchen, further strengthening the ties that explicitly bound food to self. While at the sanitarium, eating the right foods was a moral duty for upstanding behavior as much as it was an attempt to feel better physically.

Diet List No. 3—the "Nitrogenous Dietary"—was designed to stimulate the secretion of gastric juices within the stomach, and was therefore a prescription for cases of hypopepsia and apepsia, in which the production of gastric juice was found to be lower than normal. The most "stimulating" food elements were albumen and casein—albumen found in eggs and nuts, and casein found in milk and vegetables. The required foods, therefore, consisted of the following:[53]

Diet List No. 3

Beans puree
Peas puree
Nuts puree
Milk
Gluten biscuit
Gluten meal
Eggs
Beans
Peas
Lentils
Almond Meal
Lentils with nuts
Peas with nuts
Beans with nuts
Bromose ("selected and prepared nuts")
Nut porridge
Nut butter
Nut meal

One example of a specific rendering of these food options is the 1899 Prescription Number 59, intended to serve between one-half and one-fourth of a patient's daily consumption, depending on his or her state of health. Prescription Number 59 consisted of a pint of milk, a pint of Kumyss (sterilized milk mixed with lactic acids), and malted nuts. This totaled 306 calories obtained from proteids, 538 from fats, and 427 from carbohydrates, respectively, totaling 1,271 calories. This particular prescription was meant for patients with a deficiency in the production of "gastric juice," or hydrochloric acid, in the stomach.[54]

A food that commonly appears throughout the Diet Lists is a prepared bread called Granose, which was made by thoroughly cooking wheat, compressing the gruel, then baking it until it became a brown crisp. Granose appears in the Aseptic Dietary (List No. 1), the Dry Dietary (List No. 2), the Dilatation Dietary (List No. 7), and the Anti-Fat Dietary (List No. 8). As the most universally safe food to eat for any patient, Granose is particularly noteworthy because of its precursory role to the invention of Corn Flakes.

While seeming to solve the problem of fermenting bacteria inside the stomach, crispy breads like Granose were unsatisfactory in one important way: flavor. Sanitarium staff implemented stricter diets for patients who were ill, reducing the variety of foods in their meals. This meant that as visitors were diagnosed with various stomach disorders, they tended to get stuck with a steady stream of overbaked crackers. The crackers took the fun out of eating, something that did not seem to bother Kellogg: "every bit of pepper-sauce, ginger, and pepper, every piece of rich pie, every piece of cheese, sausage, ham, and all that sort of thing that goes into a man's stomach is poisoning his blood and cutting off the other end of his life."[55] In the view espoused at the sanitarium, a tasteless diet was a small price to pay for perfect digestion, stable health, an ordered society, and a balance of nature. For many patients and visitors, though, it was far too high a price, and so in Battle Creek a subeconomy emerged that provided more traditional foods. A "meat speak easy" called the Red Onion, for example, served beer, steaks, and onions to patients who snuck away for a break from the alternate cultural reality inside the walls of the sanitarium.[56] Another nearby restaurant known as the Sinner's Club operated in conjunction with the city fire department, giving refuge to sanitarium patients in search of a place to eat meat and light up cigars.[57] With a diet that precluded not only meat but also sugar, desserts, gravies, condiments, and even salt, the sanitarium chefs had—not surprisingly, perhaps—failed to discover how to make food that matched the tastes of their constituency. What remained was "a rather uninviting residue" consisting of pearl barley and cracked wheat gruels, which required overnight steaming to make them edible.[58]

After a patient was diagnosed with one of the twenty-six identified forms of digestive maladies, a sanitarium physician—often John Kellogg himself—would give an order for a diet prescription. Each of the prescriptions was made "not only with reference to the proper combinations of food substances in relation to digestion, but also in relation to the food elements which they supply." This meant that in addition to decreasing stagnation and bacteria, the meals were Kellogg's attempt at creating nutritionally balanced meals, the list of which "may be enlarged at will by the necessary chemical examination of other substances."[59] Bringing nutritional science into the experiment kitchen was a devotional act, reflecting Kellogg's belief in formalizing knowledge in all spheres, including food. The historical moment at which modern digestion rose to prominence at the sanitarium was the same moment at which the science of human food itself began a codification process, one defined by parsing out the constituent elements of nature consumed by people.

HISTORY OF NUTRITIONAL SCIENCE

The arrival of human nutritional science to the United States in the late 1880s offered John Kellogg a tool with which to buttress claims about "eating right" that he and Ella had been making since the 1870s. While by the 1880s the term *science* had already been applied to many industries and activities, the scientization of food preparation and consumption had not yet emerged as a widespread way of imagining one's actions at the dinner table. As Ella wrote, "the preparation of food has been less advanced by the results of modern researches and discoveries in chemistry and physics, than any other department of human industry."[60] She cited industries such as agriculture, iron mining, and glass making as departments that have modernized with the help of science, but food preparation and consumption were surprisingly lacking. The sanitarium would become the first major institution in the United States to direct modernist rationality toward food consumption. The inherently small scale of molecular analysis and the subsequent categorization of food's constituent parts—

namely carbohydrates, proteins, fat, and calories—gave the Kelloggs a language with which to formalize and quantify what before, in the home health book genre, had been mostly a matter of custom and faith.

When Wilbur Atwater was appointed director of the federal office of the Agricultural Experiment Stations in 1888, he would use his position to explore what was the new field of scientific human nutrition. Previous forays into the relations of food consumption and human physiology were undertaken most earnestly by the German chemist Justus von Liebig. Liebig was ultimately interested in how materials coursed around the earth and through bodies, and what effects that movement had on plants, animals, and people. Circulation of matter took on different queries for Liebig, from the constitution of blood, to the "metamorphoses of tissues," to the "restitution of an equilibrium in the soil."[61] This pairing of interests in human physiology with agriculture predicts how the field of human nutrition would emerge in the United States from an unlikely source, one that on the surface had nothing to do with humans and everything to do with crops: the Agricultural Experiment Stations. Almost all of the founders and leading practitioners at the American experiment stations trained directly with Liebig, and found ways to extend his concerns to human nutrition.[62] Liebig, who studied agriculture, physiology, and botany, saw that the twin processes of absorption and circulation found in soils and plants were applicable to the human body as well.[63] He claimed that the passage of the elements of digested food through membranes of the intestinal canal and subsequent entrance into the bloodstream operated under the same principles as the movement of nitrogen, carbon, phosphorous, and ammonia through the roots and leaves of plants into their sap. In this way, bodies shared the same properties as the physical environment. Looking to nature to explain human physiology, Liebig marks the earliest modern forays into the chemistry of human nutrition, and his efforts were brought to the United States most notably through one of his American students, Wilbur Atwater.

Training in chemistry (as well as other disciplines) at a German university differed from that in U.S. institutions in three notable ways. One was *Lernfreiheit,* or freedom of learning, meaning that the student could choose his or her faculty and lectures as deemed appropriate by the student. Second was *Lehreiheit,* or freedom of teaching. Ideally this benefited the student because he or she gained access to original ideas, also benefiting teachers by protecting them from state restraint. And third was the notion of the original research project. This was the element of German training most unlike that in the United States, because it meant that "mere acquaintance with a body of knowledge was not enough. . . . [T]he first task of the student of chemistry was to master the methods of science; he then had to apply those methods to at least one of the unsolved problems in his field."[64] All told, about a thousand American chemists received advanced degrees in Germany between 1850 and 1920.[65] Those returning would push for change in the American system, which focused on creating a college experience that would make cultured citizens rather than on what was seen in the United States as specialized technical education. This change eventually happened because the results of scientific experimentation were powerful and profitable engines to industry, and came to gather a lofty cultural status of their own.

Securing funding for his new human nutrition research was not easy, and occupied a number of years of Atwater's professional life.[66] His success rested in his ability to frame nutritional science as a prerequisite to solving the problem of industrial wages and labor productivity, an argument that counted on the conception of workers as machines.[67] Due to rising labor unrest, policy makers and social observers realized that wages in the industrial sector were too low. Rather than suggest a wage increase or admit to a broken economic system, Atwater claimed that he could scientifically formulate the most efficient dietary regime for the American worker.[68] His plan would obliterate differences not only in individual body type and eating patterns, but also among the varied regional and ethnic culinary traditions among the nation's industrial workers. The plan was based on the logic of separating food products into their constituent parts, then mathematically figuring which combination of foods would provide all the necessary elements. By 1896 he had coauthored an entire volume that listed "the chemical composition of American food materials" so that the working class could be properly instructed by the federal government on what foods would provide their machine-bodies with the exact minimum amount of energy to get through the day without ruining their entire family budget.[69] The similarity between Atwater's chart and Kellogg's food charts is not a coincidence, as Kellogg borrowed from the emerging trend in nutritional science as promoted through the Agricultural Experiment Stations.

CONCLUSION

Kellogg brought hydrotherapy into the manifold of hard science, using the microscope to build a language of medical evidence (e.g., intestinal bacteria) that would rebuke some inherited hydrotherapeutic practices while evolving others into his own. The inclusion of microbes in Kellogg's discourse on health came to distinguish the sanitarium from other water cures by attempting to bolster environmental etiology with a rhetorically more powerful scientific explanation. This likely would have led most practitioners on a path away from the old paradigm and into the new one. But Kellogg's devotion to Adventist theology made it impossible for him to jettison the spiritually inflected "nonmedical" methods that grew alongside water-cure culture. The entire program of healing at the sanitarium rested on the turn to the invisible. Kellogg exercised the tool of the microscope to maintain an intellectual upper hand over his patients, and also in attempts to demonstrate his legitimacy to the mainstream medical community—a legitimacy that he never fully received. And while he may not have talked about it with the same language, or even explicitly at all, Kellogg's turn to the small, unseen aspects of the body opened health to the visible, lived-in environment, defining an entirely different body–environment relationship than the one given to him by Jackson's vitalism and the nonnaturals.[70]

During the twenty years between Kellogg's appointment as sanitarium director and the publication of his 1896 book *The Stomach,* the stance on diet that emerged from the Seventh-day Adventist roots had transformed, going far beyond moral suggestions to become the central, scientific pillar of health. Among all the other "natural" modes of healing embraced by Kellogg—hydrotherapy, calisthenics, massage machines—eating stood out because it was the method by which digestive health could be materially and immediately affected. The inclusion of microbes in Kellogg's discourse on health came to distinguish the sanitarium from other water cures by bridging environmental and bacteriological etiologies. When

Kellogg wrote that "it seems quite improbable that any of the vital processes of the body should be dependent upon the action of destructive microbes for their perfect performance," he wore his philosophy of autointoxication on his sleeve.[71] His application of chemical analysis to the process of digestion *as well as* to the composition of foodstuffs made eating at the sanitarium a conspicuous act at the leading edge of nutrition, performed by an aristocracy of social reformers. Inventing the pharmacological use of food required someone like Kellogg, who had specialties in both gastroenterology and nutrition. Packaging these spheres into one turned out to be a wildly popular experiential product attained by staying at the sanitarium, and eventually a boxed, wax-sealed product, attained by purchasing breakfast cereals. In the early twentieth century, people sought each of these out in droves.

3 Flaked Cereal

THE MOMENT OF INVENTION

This chapter is about the moment of invention of the first flaked cereal in 1894. This moment is important for two reasons, each the marker of profound changes in the relationship between people and food in Battle Creek, specifically, and in American culture more broadly. First, when the Kelloggs invented flaked grains they introduced a mechanical prosthetic for the digestive system, outsourcing the work of the stomach, and thereby giving birth to a biotechnology that geographically extended the body. Second, and concurrently, they solidified and exported one version of American cuisine that to this day remains recognizable. This version of American cuisine is practiced by people who believe there are systemic, ambient dangers in the greater national food system, and then move to counter these problems. The spirit of this cuisine is defined less by a collection of foodstuffs (the way a cuisine is typically defined), and more by its growing and manufacture of edibles based on a reactive ideology that sees the prevailing food system as potentially poisonous. For Kellogg, this was the process of engineering solutions for a society full of bodies incapable of handling the terrestrial reality of its food system without falling into illness. Today's industrial food economy takes a page from Kellogg; the very notion that something would go horribly wrong with the entire food system without constant technical intervention means that engineering "safe" foods is lending a hand to bodies incapable of digesting food straight from nature. In using technology to correct what he saw as the imperfections of nature, Kellogg gave a cultural excuse—based in health—that justified the political economy of agriculture that is so prevalent, and so heavily critiqued, today.[1]

But the spirit of the type of cuisine Kellogg popularized goes even further, into the core of eating cultures that seek to *counter* the prevailing, industrial American food system. The political-economic gears that now turn to transform fields of grain into packaged products would have likely received mixed reviews from John Kellogg. On the one hand, he would have been skeptical of the "unnatural" ingredients found in many of these mid-twentieth through early twenty-first-century foods, like the lists of preservatives one can find on the back of so many prepackaged foods. He also would have abhorred what he deemed unhealthful ingredients like corn syrup, and a global food market typified by, for example, the mass distribution of beef. But stepping back to the bigger picture, and more significant to the point I am making, Kellogg would have been on board with the notion that there must be something very wrong with the *prevailing system* of making and distributing food for the nation. It is here where we can draw a thread through the invention of breakfast cereal, to the agro-industrial complex, all the way to the countercultural food movements. It is this style of interrupting the movement of food from field to table that Kellogg brought to prominence in Battle Creek, and that endures as definitive of American cuisine.

That Kellogg "manufactured" and "engineered" solutions to food, while at the same time found something mightily disturbing about the overall food system, highlights what we now might interpret as a paradox. These days the thrust of healthful eating advice is based on the idea that food should be manipulated as *little* as possible from how it is imagined to exist in a pure, natural state, and that this takes a great overhaul to achieve. Kellogg's story highlights that part of American cuisine is about never being satisfied with the movement of organisms from nature to the eater's body. The paradox is that we now see Kellogg's practice of manufacturing foods as part of the industrial food industry. The shared component with alternative food movements, though, is that Kellogg was as obsessed with getting food right as are large portions of the American eating culture today. This obsession with getting food right is the part that I believe defines American cuisine more than a collection of foodstuffs. Kellogg saw dangers in the prevailing food system and tried to correct them by altering the food-making structures and technologies that he inherited. Today we see dangers in the prevailing food system, too, and we try to correct them by purposely reorganizing the food-making structures and technologies that we have inherited.

To understand the birth of Granose, or wheat flakes, in 1894 it is important to understand that since at least 1877—the year after he became director of the sanitarium—John Kellogg had been continually seeking to update and refine the food selections offered to patients in Battle Creek. In his spare time he conducted experiments with grains and fruits, using mostly oven heat to modify the food's properties and textures. Six years later, as his efforts grew in complexity and intensity, the endeavor was given its own designated space—the sanitarium experiment kitchen—with Ella Kellogg appointed as its manager. The opening of the experiment kitchen marks a major shift in how health reformers approached the issue of food. John and Ella Kellogg's refusal to accept the bland—what his sanitarium patients called "meager and monotonous"—foods that had up until then defined the health reform cuisine meant that he would invest years of his working life, using hundreds of staff members and thousands of patients as guinea pigs, to make the perfect health foods.[2] In 1892, nine years after the inception of the kitchen, and after fifteen years total of food experimentation, Ella thoroughly articulated the Battle Creek way of eating up to that point. Her book *Science in the Kitchen* is the formal expression of what had been going on in the experiment kitchen, including the philosophies that she and John had adopted from other health reformers and pruned into their own set of recipes. One of those recipes was called Granola Mush, the process for which is described as follows: "Granola, a cooked preparation of wheat and oats, manufactured by the Sanitarium Food Co., makes a most appetizing and quickly-prepared breakfast dish. Into a quart of boiling water sprinkle a pint of granola. Cook for two or three minutes, and serve hot with cream."[3]

Granola—a food that still today is associated with healthfulness and simplicity—finds its origins in the water-cure establishment of J.C. Jackson in 1863. At the same time that Jackson was inspiring Ellen

G. White and her family at Our Home on the Hillside in Dansville, New York, with hydrotherapy, he was trying his hand at making new foods, as well. While Jackson did not prioritize food manufacturing over the power of water, his brief forays did leave a lasting thread that would be picked up about a decade later by the Kelloggs. By mixing coarsely ground whole-wheat Graham flour with water and then baking it, Jackson ended up with a dense lump of hard bread. He then broke the bread into pieces and baked them again at high temperature, resulting in small, hard bits of grain that he called "granula."[4] A follower of Sylvester Graham's philosophy, Jackson widely pronounced the benefits of eating without "overstimulating" the body in his publications at Dansville. The nuggets were not unlike what today one might recognize as Grape Nuts, except that they were "exceedingly firm." They were so hard, in fact, that they had to be soaked overnight in water or milk before they became chewable. Because of their unyielding, solid structure, among Jackson's Dansville guests granula quickly took on the moniker of "wheat rocks."[5]

In one of John Kellogg's early successes with food, he and his team made a derivation of Jackson's granula by adding cornmeal and oatmeal to the Graham flour before baking it for the first time. After breaking the resulting hard bread into pieces and baking them again at high temperature—just like Jackson—Kellogg ended up with a multigrain version of granula. To sidestep any serious legal ramifications, he called the food "granola."[6] When first made in Battle Creek in the late 1870s, Kellogg's granola nearly replicated Jackson's granula in content, size, taste, name, and—unfortunately—hardness, too. Upon cracking her teeth after fulfilling Dr. Kellogg's prescription to eat a plate of dry granola, one woman at the sanitarium demanded that Kellogg pay for her new dentures.[7] This mishap was, for Kellogg, the last straw in his failing attempts to make appealing health food. And so, by the turn of the final decade of the nineteenth century John Kellogg put his finger on a challenge that he had been living with for over ten years in his operations at the sanitarium: "A sanitarium must provide food prepared in such manner as to be both wholesome and palatable, tempting to the patient whose appetite is perverted and fickle, and at the same time easy of digestion and highly nourishing."[8]

In this articulation he takes on an air of accomplished superiority, as if he is telling other promising sanitarium directors to simply will wholesome, palatable, nourishing, and easily digested food into being. But this confidence disguised what had in reality become a puzzle, one that if solved would not only help Kellogg realize his vision for digestive health but also drive forward the business success of the sanitarium well into the twentieth century. In 1890 Kellogg knew the problem well enough to proselytize to others, but he continued to work on the solution himself behind closed doors. In short, this challenge was to make food that was healthy *and* tasted good; food had to be nutritious and delicious.

The Kelloggs faced many obstacles in making foods that both met their strict dietary standards and "tempted" patients' appetites. The strongest of these was undoubtedly the existing cuisine to which the sanitarium diet played foil. Nineteenth-century American eating habits have been widely recorded as gluttonous, rich in meats, fats, sugars, and starches. Food historian Andrew Smith points out, for example, that a hearty American breakfast "might have included such things as lamb chops, tripe, clams, broiled salmon, beefsteak, liver, kidneys, bacon or ham, smoked fish, codfish cakes, poultry, muffins, waffles, pancakes, potatoes, fried, boiled, or scrambled eggs, corned beef hash, mush, grits, hominy, bread, rolls, and seasoned fruit."[9]

This standard American cuisine is exactly what Sylvester Graham—a godfather of the health reform movement and one of Kellogg's inspirations—had been railing against for his entire career, promoting whole-grain diets while trying to abandon processed flours, bacon, eggs, pancakes, and syrup.[10]

SYLVESTER GRAHAM AND OVERSTIMULATION

When it came to food, though, the health reformers were far better at explaining why people should eat differently than they were at providing viable alternatives. The litany of unappealing health-reform food is traced back to the efforts of Sylvester Graham in the 1830s, whose bran flour was a response to what he saw as the moral and physical dangers of eating breads made with refined flour. "Superfine" flour, as he called it, was a new type of flour made by removing the husks of the grain before milling. This created a fine, powder-like substance that, when leavened with brewer's yeast (another transgression) and baked, resulted in something similar to what we would now call white bread. Texture was not the only problem that Graham identified in refined flour. He also exposed the "chemical agents," or additives, that bakers were using, including "alum, sulphate of zinc, sub-carbonate of magnesia, sub-carbonate of ammonia, [and] sulphate of copper."[11] When Graham launched a national public campaign rejecting refined flour, he gained traction because his message was a lightning rod for the perceived (and real) effects of commercializing the household economy. The Jacksonian period in the 1830s is characterized by historian Stephen Nissenbaum as a time when "more people were starting to buy rather than to make what they used, and to sell rather than use what they made."[12] This means that a geographically dispersed marketplace was replacing local economies, moving the production of material culture—including things like furniture, tools, and, significantly, bread—away from the home and into bigger markets. These places, the sources of consumer goods, were potentially, for the first time, completely unknown to consumers. Graham was horrified that bread had become a generic commodity, as in his view it threatened a perceived local, family-centered social fabric as well as the fibrous quality of the foodstuff itself.[13]

To further strengthen his case, Graham bolstered his social critique with ideas from medicine. A grand theorist at heart, he sought to build a system of understanding the food–body relationship wherein ideas from any existing medical scheme would ultimately prove him right. The most lasting and influential medical theorist Graham used to build his system was François Broussais, who, as historian James Whorton put it, "concluded that all disease stems from overstimulation of the body's tissues, especially those of the gastrointestinal tract."[14] This greatly appealed to Graham because it gave him reason to promote the consumption of whole-grain bread to counteract stimulation of the bowels brought on by meats, coffee, alcohol, and spices. To suppress the stimulation that arose from eating and drinking was to promote the moral godliness expressed in the Victorian antipleasure principle, but for Graham it also had a direct connection with gastrointestinal comfort, two phenomena that were not unrelated in his mind.[15] Reducing and even eradicating stimulants that could roughen up the body in any way became the driving force behind Graham's proselytizing, and *overstimulation* became his buzzword.

In addition to promoting bran bread that was unaltered "from its natural condition," Graham's removal of spices, condiments, sugar, and meat from the diet (along with all liquids save water) would greatly reduce the number of acceptable ingredients when formulating his famous Graham crackers and other health foods.[16] What would become a buffet of health-reform foods later in the nineteenth century was, by definition, born tasting bland. It was *because* the foods lacked flavor and richness that they promised to save American society and make people feel better. Thanks to the relentless preaching of Graham and his contemporaries—notably William Alcott—in the 1830s and 1840s, the dry and boring but morally and medically right foods gained some popular momentum. The legacy of overstimulation that Graham popularized set a baseline for what his intellectual descendants would use to concoct more health foods.

GRANOLA: NOT QUITE GOOD ENOUGH

Just two years after John Kellogg proclaimed that sanitarium food must be palatable and healthful at the same time, Ella Kellogg published her recipe for Granola Mush, doubtlessly the result of significant tinkering that sought to soften the nuggets following the cracked teeth incident. By the early 1890s all gears in the sanitarium engine were turning toward finding new foods that aided digestion and were attractive to a wide range of patients. However, the reason that Granola Mush required only a few minutes of soaking to soften up (as opposed to Jackson's granula, which required an overnight soak) remains something of a mystery. It is possible that the Kelloggs employed a convenient linguistic fabrication—*mush*—to demonstrate progress toward a perfect food. They seemed to be grasping for anything that would appear to diverge from the existing state of culinary affairs, anything that could be touted as a marked improvement from dry, hard, tasteless bread kernels. Granola, then, for all of its disagreeable features, remained on paper a food that approached perfection when it came to corporeal health.

If eating was divorced from taste and pleasure (as the health reformers undoubtedly wanted it to be) and was instead a calculable, instrumental activity used to make bodies right (as Kellogg undoubtedly wanted it to be), then in 1893, a year before the first wheat flake, granola was the best thing one could eat.[17] Therefore, to understand why the manufacture of a light, crunchy flake was seen as such a great advance, it is first imperative to understand why the Kelloggs thought so highly of granola in the first place. As early as 1889, and into the 1890s, the Kelloggs shipped out boxes of their granola at the clip of two tons per week to former patients under the banner of the Battle Creek Sanitarium Health Food Company.[18]

Even more so than his science-based departure from Graham's school of health reform philosophies, John Kellogg's headlong charge toward a new era of health foods was guided by how he made sense of the incorporation of food into the body by the digestive system. Making the perfect food was a matter of making it perfect for digestion. As Kellogg explained it, "A dry crust of bread, chewed for some minutes, becomes sweet. . . . In acting upon the starch, the saliva produces first soluble starch, then dextrin. . . . This conversion into sugar constitutes the digestion of starch. *It is essential, however, that the starch should be cooked, as the saliva cannot digest raw starch.*"[19]

The conversion of starch (a carbohydrate) into dextrin (a translucent, adhesive, gum-like substance) was a chemical process known to be the first step in the human digestion of grains. In part because it fit so nicely with the theory of auto-intoxication, for Kellogg it served as the single most explanatory observation concerning the relationship between food and the body. Specifically, Kellogg believed that the human digestive system was not built to efficiently convert grains that had not been predextrinized. By making the stomach work harder, the foods would stay there and in the remainder of the digestive system for too long, rotting, festering, and causing myriad illnesses. This system of understanding is hinted at in an 1894 advertisement for Granola, which states that the grains were "treated" so that their "irritating character" had been eliminated (see figure 11). Nature had not set up the food–body system to streamline incorporation. Instead, that task was left to human technological ingenuity, which Kellogg took on like a possessed visionary.

Granola was popular enough, but John Kellogg wanted something more. He did not want more promotions or sales, but instead wanted a food that was perfectly dextrinized, easy to chew and swallow, and he wanted to be explicit about this in his lectures and advertisements. He felt like he was not performing his sacred duty to cure the nation by shipping

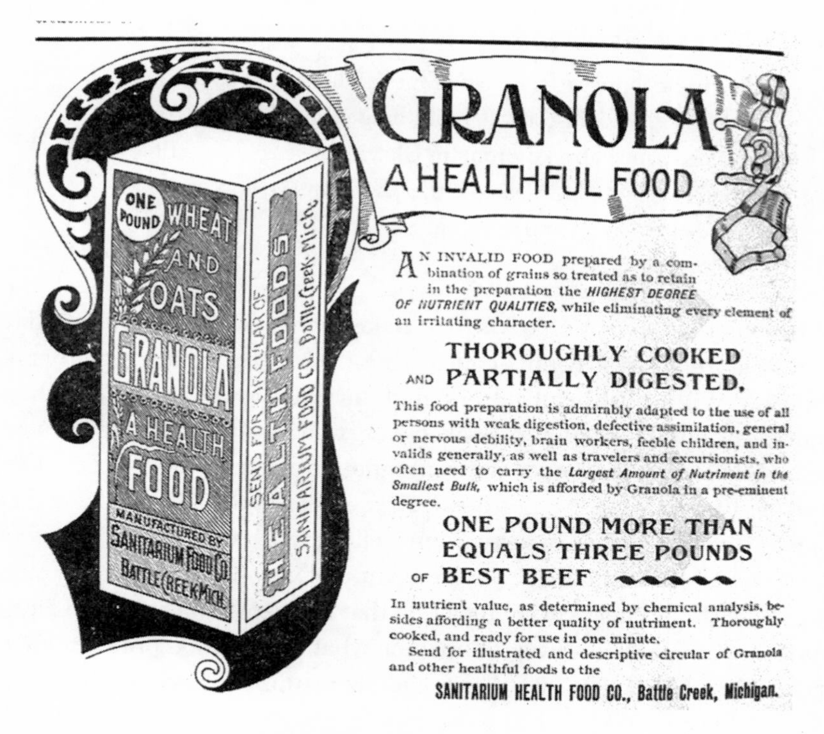

Figure 11. An advertisement for Granola, in which its ability to ease digestion is suggested with the phrases "eliminating every element of an irritating character," "partially digested," and "adapted to the use of all persons with weak digestion." Behind these suggestions was John Kellogg's adoption of the theory of auto-intoxication. This rhetoric would only increase with the advent of flaked cereals. Image reproduced from *Modern Medicine and Bacteriological Review* 3, no. 2 (February 1894). Courtesy of the Bentley Historical Library, University of Michigan.

a food product when he knew there must exist a superior one, just out of reach, just beyond the clouded veil of invention. He wanted a food that seamlessly integrated with the physiology of the digestive system, almost as if the stomach itself did not know that anything occupied its chamber because it worked so little to assimilate the contents. The key to making this food would rest in the absolute conversion of starch into dextrin *before* food touched the inside of the mouth and was advanced to the stomach.

THE MOMENT OF INVENTION

> The original purpose in making the toasted cereal flake was to displace the half-cooked, pasty, dyspepsia-producing breakfast mush, by a thoroughly cooked and easily digestible cereal which might be eaten either dry or moist, and which would enter easily into solution.
> —John Kellogg, 1908[20]

The moment of invention of the first flaked cereal in 1894 has accrued its own folklore. Most accounts concur that both of the Kellogg brothers—John and Will—were involved, and that the invention could not have happened without a container of cooked wheat being left out in the open air, seemingly forgotten, for a significant amount of time. The context of the story is that, in making pre-dextrinized grain foods, the Kelloggs had encountered a creative impasse. John Kellogg knew that he wanted a food that was, first, free from the vexing texture of granola, and second, more aligned with the chemical composition that he deemed beneficial to digestion. He struggled, however, to imagine what type of food product would satisfy these demands—its shape and size—and furthermore how that food could be produced.

The impasse occurred because the Kelloggs at that time were at the end of a path. It was a well-established path that had been trod by two generations of health reformers who preceded them, led by Sylvester Graham and J. C. Jackson. It was Jackson's granula, derived from Graham's whole-wheat flour, that formed the basis for Kellogg's granola. John Kellogg was, as historian of science Thomas Kuhn might have put it, trapped in a paradigm—a particular way of thinking and interacting with the ideas and technologies available to him, the "cumulative sedimentation of theoretical systems and empirical materials."[21] The culinary arm of the health reform movement had built up a paradigm in which all of its constituent parts seemed to be immovable. To make healthy food one had to use whole-wheat bran flour, had to mix this flour with water and/or yeast in various proportions, had to bake it for varying lengths and at varying temperatures, and had to break it up into various sizes. In addition to granola, most of the bread products made at the Sanitas Food Company and the Battle Creek Sanitarium Health Food Company were very similar to one another—they failed to escape the paradigm of "meager and monotonous."[22] Even after years of work in the experiment kitchen most sanitarium bread foods shared three characteristics: (1) they were made from whole grains, (2) they were free from "unwholesome" baking materials such as lard, soda, and baking powder, and (3) they were subjected to high-temperature baking.[23]

Having returned from a visit to fellow future cereal magnate Henry Perky's Shredded Wheat factory in Denver in 1893, John Kellogg became utterly convinced that granola had become passé, and that there was a way to make foods even more prepared for digestion.[24] The expectation of high taste that guests increasingly brought to the sanitarium in the 1890s was beginning to clearly highlight a contradiction: the mission of a purported hypermodern health facility was at odds with the practice of serving Grahamite food invented sixty years prior. To address this almost embarrassing shortfall, John Kellogg went back to the drawing board, so to speak, with greater ambition than ever before. He began imagining ways to combine the shape and texture of zwieback (softer, leavened toast slices) with the small size of the granola bits, and at the same time completely predextrinize the food. John and Will Kellogg began looking for a way to turn "each grain of wheat into a small flake of toast."[25] Doing so would solve the texture problem and the chemical-content problem at the same time.

Making individual, non-crumbling, flat, chewable, miniature pieces of toast might not seem an exceedingly difficult thing to do. But when the Kellogg brothers began their attempts, problems ensued. In the earliest trials they soaked raw wheat overnight in water, then fed the resulting mass through the pair of steel rollers that they had been using for years to smash pieces of hard-baked bread into granola. Predictably, this resulted in not much more than puddles of watery starch along with globs of coarse bran. Lightly steaming the wheat mass before feeding it through the rollers did not help much, as it transformed into a paste-like, dense aggregation that clogged the rollers.[26]

The challenges that the Kelloggs faced in identifying a process for manufacturing pieces of tiny toast persisted until—perhaps fueled by frustration and futility—an accident of forgetfulness occurred. Crawling out from under the "cumulative sedimentation" of habits and materials that constituted the health reform food paradigm was made possible because someone accidentally left a batch of water-soaked, boiled wheat sitting on the kitchen counter for several hours before putting it through the rollers.[27] There was no prescription for this step in the process—no inherited custom of food making that would have caused the Kellogg brothers to wonder if, just maybe, they should leave the wheat mass sitting uncovered in the open air before trying to roll it into bits. Nothing could have introduced this step into the empirics of experimentation save a complete—and, in this case, lucky—accident. When John Kellogg returned to the chunk of what was now drier, moldy wheat, he decided that rather than discard what he thought would surely be a wasted trial, he would roll it through anyway. As his biographer Richard Schwarz tells it, "He fed the wheat into a small pair of rollers while one of his foster children turned the crank and another used a large bread knife to scrape off the distinct flakes that emerged. Success had come. Unknowingly, the doctor had stumbled upon the principle of 'tempering,' a process basic to the future of the flaked cereal industry."[28]

Horace Powell, the biographer of Will Kellogg, tells the story differently. In his version it was days—not hours—that lapsed between the boiling of the soaked wheat and its rediscovery, and Will played a much more prominent role. According to Powell, "it was Will Kellogg's duty to squat down underneath the rolls and scrape . . . by then, the cooked wheat had become decidedly moldy, the two brothers decided to run it through the rollers to see what would happen. Much to their surprise, it came out in the form of large, thin flakes, each individual wheat berry forming one flake!"[29]

However it exactly happened, by soaking the raw wheat, then boiling it, then allowing the mass to sit out in the tepid air of the sanitarium kitchen for some time, the Kellogg brothers had introduced a step into the process of making health reform foods that opened the possibility for making flaked cereal. This step, as Schwarz points out, is now called tempering, known to present-day food chemists as the "period during which the cooked grain mass or cereal pellets are held in collection bins to allow the equilibration of moisture within and among the particles."[30] Because they forgot the wheat mass on the countertop of the sanitarium's kitchen for an extended period of time, a rudimentary version of tempering occurred naturally. The wheat became drier, which left it more susceptible to flaking off as it went through the rollers. Additionally, what moisture did remain was more evenly distributed throughout the mass, preventing any awkward globs of over- or underhydrated wheat to reduce the uniform falling of flakes as they were scraped off the rollers with a knife. With a little more experimentation, they were able to eliminate the mold by tempering the steamed wheat in clean metal bins instead of on the kitchen counter. Once the wheat was scraped off the rollers, the still-moist flakes were baked until crisp, and Granose, the immediate precursor to the world's most well-known breakfast cereal, was born.

THE FIRST PREDIGESTING MACHINE

John Kellogg had identified nature's deficit—that is, the difference between its organic nutriment and the needs of the body—through theoretical medical reasoning coupled with chemical analysis. Balancing this deficit was a matter of enhancing nature with technology. For Kellogg the corollary of enhancing nature with technology was outsourcing the work required of the digestive system to machines that existed on the manufacturing floors of the Sanitarium Health Food Company in Battle Creek. The problem was that grains did not come from grasses ready for the body to digest. The entire digestive apparatus, from the mouth to the colon, had to work especially hard to convert starch into dextrin. When it did so, Kellogg's entire schedule of digesting, defecating, and cleansing—in short, his path to health—was derailed. At the sanitarium, as soon as food touched the mouth, the clock was ticking. Food had to meet the stomach ready to be immediately incorporated, and there was no better way to achieve this than by inventing and utilizing a predigesting machine.

Granose was not originally invented as a breakfast cereal, but as a new health food in the arsenal of John Kellogg's diet prescriptions. He began using it immediately to replace or supplement the indurate granola, and the patients on whom Kellogg's program of "biologic living" prevailed in Battle Creek started to mail-order Granose in high quantity when they arrived back home.[31] Supplying the sanitarium with wheat flakes and maintaining a bustling mail-order business required robust factory and warehouse installations. The popularity of the flaked cereals soon caught on in Battle Creek, as imitators were quick to make their own renditions and more aggressively market them in and outside of Michigan. Knowing the strength of this competition led John Kellogg to seek protection of the intellectual property by obtaining a U.S. patent.

In 1895 he filed an application for a patent titled Flaked Cereals and Process of Preparing Same, which was accepted and published in 1896. As he put it, the object of the "improved alimentary product" was to "provide a food product which is in a proper condition to be readily digested without any preliminary cooking or heating operation, and which is highly nutritive and of an agreeable taste, thus affording a food product particularly well adapted for sick and convalescent persons."[32]

In the year between the first accidental tempering of wheat in the sanitarium's kitchen and the filing of the patent application, the Kellogg brothers worked furiously to perfect the process and scale it to larger quantities. John would "jot down memoranda suggesting food experiments" and send them to Will, who worked up to 120 hours per week during that time, testing all conceivable variables in the steaming, tempering, rolling, and roasting processes. For example, Will first used a chisel to knock the tempered wheat from the rollers by hand, only to replace the implement after a short time with a paper-cutting knife from the Adventist printing offices (see figure 12). Eventually he affixed the paper-cutting knife to the roller apparatus, creating a prototype for an automatic flaker.[33]

Once the process had been formalized, John Kellogg spelled out the exact steps in making their flaked cereals in the 1896 patent. It is worth paraphrasing them here, as they shed light on how the new technology was spatially extending the digestive system into the manufacturing spaces of the sanitarium grounds. In this document we witness clear signs of the hybrid machine-body coming into form that would characterize the beginning of the geography of digestion in Battle Creek.

1. Soak the grain for 8 to 12 hours, "thus securing a preliminary digestion by aid of cerealin, a starch-digesting organic ferment."
2. Cook the grain thoroughly by boiling it in water for an hour, "to the stage when all the starch is hydrated. If not thus thoroughly cooked, the product is unfit for digestion and practically worthless."
3. "After steaming the grain is cooled and partially dried [tempered], then passed through cold rollers, from which it is removed by means

Figure 12. Drawing of William Kellogg scraping flakes of wheat from the rollers. Originally used to crush hardened bread in the making of granola, the rollers became an intentional extension of the digestive system in Battle Creek. Image reproduced from Horace B. Powell, *The Original Has This Signature—W. K. Kellogg: The Story of a Pioneer in Industry and Philanthropy* (Battle Creek, Mich.: W. K. Kellogg Foundation, 1989; originally published in 1956), 91.

of carefully-adjusted scrapers." This process rolls the grain "into extremely thin flakes . . . made readily accessible to the cooking process to which it is to be subsequently subjected *and to the action of the digestive fluids when eaten.*"

4. The flakes are collected in trays, then steamed, baked, and roasted in an oven "until dry and crisp."

In John Kellogg's analysis following this four-step process, he describes the resulting flakes as "not brittle," assuming a "sweet flavor" from the conversion of starch into dextrin. He also calls the product "perfectly sterilized," noting (erroneously) that the flakes will keep indefinitely. Together, the cooking, rolling, and roasting bring the flaked grain "into a condition in which it is readily soluble and digestible without any further cooking."[34]

Though never stated, in this document we see the Kelloggs giving nature—the biological process of human digestion—a new name. Flaked cereals were much more than miniature, thin pieces of toast, they were the material assemblage of particular kitchen technologies woven together to remove digestion from inside the body. If in 1896 patients came to the Battle Creek Sanitarium assuming their bodies were constituted from elements found in nature, they must have left believing they were now part of a wild biotechnological future in which eating Granose far exceeded the benefits gained from eating mere plants. They seemed all too willing to trade in their own digestive organs for the rollers, happily extending their bodies into the network of the sanitarium that the Kelloggs were continuing to build.

THE MACHINE-BODY

Not unlike Sylvester Graham, John Kellogg believed that vegetable foods existing closest to nature promised the greatest curative potential. But unlike Graham, whose ideas align with the health food sensibilities prevalent in the early twenty-first century—a mistrust of modified, processed, manufactured, industrially grown, and shipped foods—Kellogg's was a sensibility *founded* on the promise that technology could make better foods than nature could.[35]

Overcoming the shortfalls of nature such that the digestive system was met with just the right material nourishment meant modifying raw plants with machines. By the time Kellogg was its major spokesperson in the early twentieth century, the health reform tradition that began with Sylvester Graham's nostalgia for non-engineered food is the very same tradition that engineered flaked cereals with industrial machinery. This great paradox of the health reform movement's philosophy concerning food—and the wider public's willingness to accept the paradox—allowed Kellogg to endorse the inherent healing power of natural foodstuffs while simultaneously proclaiming that foods taken directly from nature left the modern human body wanting. After an experiment conducted in 1895, for example, he found that, of 4,875 stomach samples (from patients whose stomachs had been pumped after a test meal), starch conversion had only occurred in 669. He concluded that "starch, in order to be digested by the saliva, must be cooked, and the more thoroughly it is cooked, the better."[36] As he had already experienced for almost twenty years, though, it was not enough to blast a grain concoction with heat and assume that it would be palatable. While this approach gave fantastic lab results based on analyzing the contents of a patient's stomach that had been extracted with a tube, it only fulfilled the first half of the "nutritious and delicious" equation.

The 1895 advertisement shown in figure 13 most succinctly describes the benefits of Granose as a foodstuff that encapsulates an early version of a techno-natural hybrid. The copy reads, "By the use of special machinery, the wheat is brought into the form of delicate flakes . . . the starch largely converted into dextrine [*sic*], and thus made ready for solution by the digestive juice and for prompt assimilation." The special machinery to which the text of the advertisement refers are the two 8-by-24-inch rollers that had been repurposed from the granola production line (see figure 12). With the invention of Granose, an important phenomenon occurred concerning the use of machinery in the culinary arts. Namely, the relationship between technology and bodies was redefined such that the body was no longer an independent organism. It would be a premodern nightmare for Kellogg—"savage" was his exact word—for people to roam the earth eating plants and animals from its bounties without tempering the organisms first.[37] It is important to note that this is precisely opposite the idea of "whole foods." The grocery store chain that uses this name founds its entire business model on the notion that customers have access to food that is most directly from nature. The use of rollers to make flaked cereal was the first consciously applied prosthetic for the digestive system, the first intentional geographical removal of an alimentary organ function to another location. This removal is the very type of technological intervention against which whole foods play foil.

At the sanitarium, this external location was the experiment kitchen in the basement of the building, as well as the Sanitas Food Company's factory, an adjacent building on the same property. As production grew and national sales increased, the steamers, tempering bins, rollers, and ovens became nodes in the cyborg-like functioning of digestive systems throughout the continent. In 1895, the first full year of Granose production, the Kelloggs sold nearly sixty tons of wheat flakes, making Battle Creek the figurative and material center for an efficient corporeal metabolism.[38]

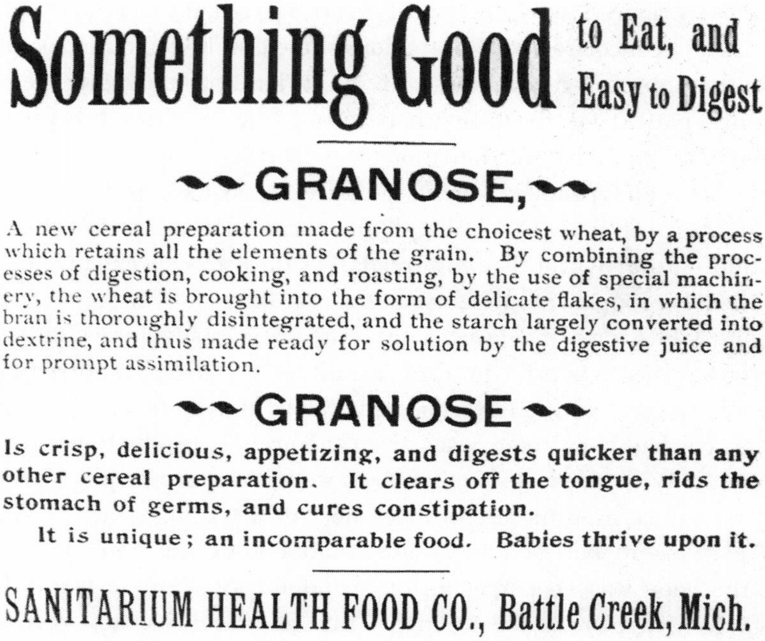

Something Good to Eat, and Easy to Digest

GRANOSE,

A new cereal preparation made from the choicest wheat, by a process which retains all the elements of the grain. By combining the processes of digestion, cooking, and roasting, by the use of special machinery, the wheat is brought into the form of delicate flakes, in which the bran is thoroughly disintegrated, and the starch largely converted into dextrine, and thus made ready for solution by the digestive juice and for prompt assimilation.

GRANOSE

Is crisp, delicious, appetizing, and digests quicker than any other cereal preparation. It clears off the tongue, rids the stomach of germs, and cures constipation.

It is unique; an incomparable food. Babies thrive upon it.

SANITARIUM HEALTH FOOD CO., Battle Creek, Mich.

Figure 13. An advertisement for wheat flakes, or Granose, the immediate precursor to Corn Flakes. Image reproduced from the *Bulletin of the American Medical Temperance Association,* 3, no. 1 (November 1895). Courtesy of the Bentley Historical Library, University of Michigan.

The invention of Granose signaled the twin arrivals of a mechanical, spatially extended body and a new food item that would eventually change the breakfast meal throughout the world. These two phenomena should not be conceived as separate or coincidental, and the key to understanding why rests in the very notion of *cuisine* itself.

CUISINE

Part of what makes *cuisine* the right concept for understanding the program of eating at the sanitarium in the 1890s is the encompassing nature of the word. Cuisine is more than a unique assortment of organisms from nature (i.e., food items) all brought together. It is, of course, partly this: an assortment that by definition has a spatial component when one considers the proximal availability of plants and animals to the people eating them. But there is another, more subtle, geographical component to the concept of cuisine that concerns technique. An example will help to illustrate this. Piles of eggplants, lemongrass, and ginger, along with containers of coconut milk and fish sauce, do not make a Thai curry by themselves. There is a way to prepare these ingredients that—while it no doubt varies from person to person—consists of general steps required for the success of the dish. One must, for example, cut the eggplant first before adding it to the coconut milk or it will not cook; it will be a big raw eggplant sitting in a puddle of warm liquid. Therefore, carrying out, or performing, technique requires tools and the knowledge of how to use them such that the food ingredients come together in the desired manner.[39] The assemblage of food, tools, and knowledge at the place where cooking happens is where cuisine is made.

It is the *confluence* of food items, knowledge, tools, and technique that makes cuisine, not just the food items themselves. Why is this geographical? Because—and this is the deceivingly simple and subtle point—the confluence must happen somewhere. It must occur "in place," a category of human experience that effortlessly rolls off the tongue without thinking, and yet has a long history of critical engagement by geographically minded scholars.[40] One of the insights that has come out of these investigations into the meaning of *place* is that places act as bundlers, or looms, for various material artifacts of everyday life.[41] If part of what makes a cuisine rests in its "kitchen arrangements"—as the *Oxford English Dictionary* puts it—then performing cuisine is a decidedly geographical act, for it is the very arrangement of kitchen tools (easily reached, configured correctly in relation to one another, stored in cabinets for daily use) that makes cuisine building and maintenance possible over the course of decades. The close and intentional arrangement of tools and ingredients *in situ*, or "in place," is what makes a kitchen. This observation did not escape Ella Kellogg. When she published her 1892 *Science in the Kitchen* she devoted an entire chapter to describing how the ideal, healthful kitchen should be organized. In that chapter, for example, Ella writes that

> where for convenience it is desirable to have some provision for supplies and utensils near the range and baking table, a wall cabinet offers a most convenient arrangement . . . with doors to exclude the dust, shelves on which to keep tin cans filled with rice, oatmeal, cracked wheat, and other grains; glass jars of raisins, sugar, citron, cornstarch, etc; hooks on which may hang the measures, egg-beater, potato masher, and such frequently needed utensils; and with drawers for paring knives, spoons, and similar articles, the wall cabinet becomes a *multum in parvo* of convenience which would greatly facilitate work in many households.[42]

Here Ella describes how one should organize the materials of a kitchen to maximize cleanliness and efficiency. The subsequent insight is that the kitchen is the cuisine itself, serving as a loom, weaving together not only the ingredients and tools, but the knowledge and technique that people bring to them.

In short, a cuisine is made in large part thanks to the design of the kitchen. The problem is that the English language separates cuisine from kitchen—where cuisine is the displaced art of assembling and transforming organisms to make food, and kitchen is the grounded place where such activity actually happens.[43] But this separation is not so readily made in many other languages, where *küche* (German), *cucina* (Italian), *cocina* (Spanish), and *cozinha* (Portuguese) each denote simultaneously the art of cooking and the place where it occurs. The everyday act of cooking is subsumed in these languages' rendition of what it means to live with a cuisine, whereas in English the idea and the geographic reality tend to be divorced.

All of this starts to bear weight when we look at how operations at the sanitarium's experiment kitchen worked. To truly invent a cuisine meant that the Kelloggs had to invent and implement tools and techniques in the place of the kitchen. Ella Kellogg goes on to state in her treatise, for example, that, "For ordinary kitchen uses, small tables of suitable height on easy-rolling casters, and with zinc tops, are the most convenient and most easily kept clean. It is quite well that they be made without drawers, which are too apt to become receptacles for a heterogeneous mass of rubbish. . . . [It is] desireable to have some handy place for keeping articles which are frequently required for use."[44]

In addition to the arrangement of the kitchen itself, John Kellogg established an infrastructure that allowed him to jot and whisk away notes on a whim to his wife, brother, and other colleagues working in the experiment kitchen, and to receive news of the experiment's results within a day or two. The notes were their brand of information technology—coding, responding to, and communicating what happened among the mechanical apparatuses, the foods, and ultimately the bodies of the sanitarium guests. One prophetic 1898 note from John Kellogg to his brother, Will, who was working in the experiment kitchen, concerned sweet corn. It said, "W. K.—By all means try it. I would like to try some further experiments with sweet corn. Here is one: Soak for one, two, or three hours, as long as necessary, in a weak solution of warm sal-soda, then see that the skins can be rubbed off. Wash thoroughly in water and boil. Show me the result. J. H. K."[45]

Notes like this connected the knowledge, tools, foods, and digestive systems at the sanitarium operation.

CONCLUSION: THE KELLOGGS AND A CREATION STORY OF AMERICAN CUISINE

It is no surprise that health reformers struggled to make their new foods taste good. Up to that point, after all, the evolution of cuisines tended to proceed with organisms from nature that were available and proximal, or were directly imported (e.g., sugar, tomatoes, and tea), suggesting that generations of trial and error played a large part in defining the parameters of taste for any particular cuisine. Today when we think of "national" cuisines—for example, Chinese or Italian—and the subcuisines thereof, there are constellations of ingredients and techniques that are known to give the cuisine its defining characteristics. These attributes, while by all means flexible and accretive, are nonetheless derived from inherited customs.[46] The idea that an entire cuisine could be conceived and produced from scratch virtually overnight—and taste good—was unprecedented on a large scale, and yet that is exactly what the health reformers were trying to do—inspired initially by religious morals and then, with John Kellogg, scientific reasoning.

The cuisine of the health reformers, though, is novel as much because of the foods it created as is because of its creation story. Its foods are well enough known: baked, whole-grain breads and crackers with mashed fruits and nuts—and its omissions are possibly even more well known: no meat, alcohol, caffeine, or tobacco; no sugar; and no seasonings. These foods (and exclusions) are the recognizable objects that visually and materially separate health reform cuisine from other modes of eating. Looking back over more than a century, we can point out that consuming the indurate breads of the health reformers was a departure from what anyone at that time was accustomed to regularly eating.

Marking this cuisine as new, however, is not quite as simple as pointing out a collection of new food items. To convincingly denote the culinary activities of the health reformers as new is a matter of looking deeply into the very *process* of how the foods emerged. The creation story of the health reform cuisine is not a gradual one, wherein an ecosystem of acquired kitchen tendencies were one day written down and codified.[47] The story is, in fact, the exact opposite. Most of the food items in the health reform cuisine emerged as the result of a particular vision of what food *should* be. As a result of religious and pseudoscientific reasoning about the body (remember "overstimulation"), the health reformers unleashed their own etiology of disease as the beginning of a chain of reasoning that ended with carefully designed foods. To divorce their ideal, healthy body from the ocean of horrors found in customary mid-nineteenth-century methods of food preparation meant creating an alternative cuisine. Because they had already preconceived notions about the ideal relationship between bodies and food, health reformers drove people away from "hearty" American meals toward a style of eating that made perfect sense on paper. Writing about "improper" and "unhealthy" ways of eating, for example, J.C. Jackson said in 1871 that the American diet is made up of "food which in its very nature is calculated to produce diseased conditions of their bodies,—food whose *effects cannot* be otherwise than to produce diseased conditions of their bodies."[48] This was a public health problem. By simply participating in the everyday American diet, people were exposing themselves to grievous diseases ranging, in Jackson's words, from "dyspepsia in all its protean forms," to "torpidity of liver," "typhoid fever," and "consumption of the bowels." His critique does not go without the suggestion of a solution. Continuing, he reasoned that, "if these diseases owe their origin, as they often do, to the use of bad foods . . . if food and these other agents were properly used, persons would not have these diseases."[49] Jackson's conception of food as something that one could "properly use" reflects an instrumentality, articulating the essence of a culinary creation story where ideas come before the foods themselves, where culture trumps nature.

But Jackson's pharmacological use of food was more in line with Graham's philosophy, which held that food was corrupted by technology, its capacity to restore directly correlated with its proximity to nature. While Kellogg inherited the notion that food could be used precisely and instrumentally—like a medicine—he would virtually reverse the stance on technological intervention. For Kellogg, making foods that cured diseases meant engineering them, as if the foods themselves were the fine tools of a watchmaker.

If Sylvester Graham was the grandfather of the health reform cuisine, then John Kellogg was its king. The Battle Creek Sanitarium has even been called "the final reincarnation of the Graham boardinghouse."[50] More than any of his predecessors, Kellogg captured the culinary-arts spirit of the health reform movement and thrust it into a grander, more popular, farther-reaching orbit. What Kellogg, his wife Ella, his brother Will, and their staff did in the sanitarium's experiment kitchen starting in 1883 amplified the finite momentum of J.C. Jackson, Thomas Low Nichols, and others to such a degree that what was before a small and disjointed group of food items became an organized system of interchangeable dishes, allowing one to eat wholly within the strictures of the cuisine on a daily basis.[51]

Again, this was not an established cuisine, but one that had been built during the 1880s and 1890s by the Kelloggs. By the turn of the century, while extreme "invalids" were still assigned courses based on their particular diagnoses, patients in the general population of the sanitarium were presented with menus that changed daily, the results of two decades of food experimentation. The variety of plates offered in the dining hall of the sanitarium—including the seemingly innocuous flaked cereal—signal a change in the way health reformers thought about the relationship among nature, bodies, and technology. John Kellogg saw the body as incapable of handling untempered grains, whereas his predecessors defined health foods as those that were closest to the state of nature.

The move from granola to flaked cereal marks a bigger paradigm shift from Graham's back-to-nature foods toward Kellogg's to-the-future foods. When Kellogg wrote the patent for flaked cereals, he explicated how technology would replace nature (i.e., the interface between the stomach and plant matter). The shift therefore also defined a new way of understanding what the body was. With Graham, the body was a direct offspring of the earth's biological and chemical processes. It was what we might call today part of an ecological system. For Graham, the machines of industry only interfered with this pure connection between body and environment. When Kellogg spelled out how kitchen machinery altered digestion for the better, though, he reset the culture of food consumption for generations such that *technological interference became valued as desirable,* attractive, and even necessary in making healthy bodies.[52] For Kellogg and other food manufacturers, this conveniently aligned with the political economy of agriculture that was becoming increasingly prepared to provide food-making machines with loads of fodder.

Looking for the identity of American cuisine, it is often mistakenly assumed that one should look for food items. To be learned from the Battle Creek Sanitarium, though, is that one idea of American cuisine can be better conceived by what I think of as Kellogg's impulse—that is, the impulse to lead with *ideas about* food rather than a specific set of foods themselves. The invention and constant reinvention of what is the best thing to eat has, in the American context, been guided by a mechanical, rational formula based in the calculation of fats, carbohydrates, proteins, and calories since those categories of nutrition were first employed. To measure food's elements and match them with precision to one's body is a definition of health that began in earnest in the American context at the Battle Creek Sanitarium, and is one compelling way to understand what it means to "eat American." In this conception, American cuisine is the search for how to best avoid the perils of dangerous foods that do not most logically and efficiently generate the healthiest bodies. To re-emphasize, though, the foods—and even the rationales for choosing the foods—have changed over time. For example, today there has been an almost complete overhaul of Kellogg's "no-biotic" principle to the *probiotics* that promise to cultivate healthy intestines.[53]

While the Corn Flakes consumed in the early twenty-first century may seem far from such a profound transition marker in the history of

American eating, the moment of their invention signifies a legacy of conscious technological intervention, an added step between agriculture and eating that was guided by deep ideas about how bodies functioned. What it meant to "eat American" moved in the late nineteenth century toward a series of progressively more enhanced foodstuffs in the twentieth century, from prepared Spam and frozen concentrated orange juice to enriched breads and vitamin supplements.[54]

The type of intervention John Kellogg introduced was technological, and was based on a belief that the digestive system was incapable of a direct encounter with the natural world. The body, he believed, had been limping along for the duration of the existence of the species, and only with the insights of late nineteenth-century medical science was it finally able to be rectified. For Kellogg, the body's true state of nature was achieved only by an analytical appraisal of its hidden processes, and a subsequent outsourcing of those processes to machines. In performing this outsourcing, he revolutionized the process of making food by revolutionizing how food manufacturers could move from an *idea* about the body to the making of the food itself. Ironically (given how deeply ingrained he was in Adventist theology), Kellogg's foods were wholly cerebral, based completely on empirical evidence from thousands of tests from his patients. This is a departure from Graham, who animatedly pointed to the moral and social underpinnings of his eating philosophy. To improve nature's alimentary bounty with technology means that one must begin with an idea about how food interfaces with the body. By looking directly to the digestive system, John Kellogg did exactly this. Based on controlled studies, he first built an idea about the body's deficiencies, then sought to build perfect foods that would perfectly minimize those deficiencies. It was a mechanistic approach that—not surprisingly—relied on actual machines to carry out the vision. This is the pattern of thinking that would flourish into the twentieth century as the birth of nutritional science offered possibilities to commoditize each and every aspect of natural elements found in food. When we encounter foods that promise to fill a deficiency in our helpless bodies, we are truly participating in American cuisine.

4 Extending the Digestive System into the Urban Landscape

EXTENSIBILITY

Near the end of 1898, John Kellogg filed an application for a sewer connection with the Battle Creek Board of Public Works. When approved and installed, the sewer line harnessed the quirky intestinal ecologies being produced at the sanitarium with the built environment of Battle Creek's urban landscape. Solid fecal matter—the result of Kellogg's digestive philosophy of health—had up until then been disposed of in privies, or outhouses. Sanitarium staff would carry the human waste via a system of bedpans to sites near the sanitarium property for removal by scavenger-farmers looking for fertilizer, who would integrate the waste into their fields. But with the arrival of the sewer connection, solid waste became effluent, mixed with water and the rest of the city's sewage. It was transported far from the sanitarium grounds into the Kalamazoo River, where it flowed west to Lake Michigan, then on to the Atlantic Ocean through the hydrologic system of the Great Lakes and the Saint Lawrence River. The shift between these two different geographies of human waste was in fact a change in the geography of the bodies being treated. The claim of this chapter is that urban infrastructural changes affected the way the digestive system worked, as they were both part of the same "assemblage" of digestion.[1]

When bodies became associated with new places (e.g., the Great Lakes waterway instead of local Battle Creek soil) they also became associated with different *types* of landscapes. The language of science that overhauled the "backward" stomach into the clean, efficient, modern stomach was the same language that would overhaul the system of human waste disposal from vaults buried in the ground—what one public health officer termed "relics of barbarism"—to the invisible, odorless, sealed tubes of urban sewerage.[2] Because the bodies being made at the sanitarium depended on this technology, the technology became a part of their bodies. Much more than a simple "association," the urban infrastructure materially co-constituted their bodies.[3] The iron pipes of the sewer system were the modern, spatial extension of the fleshy, intestinal tubes of the body, fitted together almost seamlessly with flush toilets.

When I think of this relationship between landscape and body, I consider it to be something more than "relational." Building city sewer systems in the nineteenth century led to greater sanitation and greater efficiency in human waste disposal, the effect of which was fewer intestinal illnesses and better overall health among those populations. The problem with using a metaphor of relation in this scenario—as environmental thinkers have consistently done—is that it continues to ingrain the separateness of landscape and body.[4] One causes the other to be effected in some way; the tail of landscape wags the dog of the body, or vice versa. Instead, for John Kellogg to fabricate a new type of human stomach, he had to enter a paradigm in which body and landscape became one thing: body-landscape.[5]

Looking back on Kellogg's practice, we learn that hybrids, and the entire notion of the biological merging with the technological, become normalized in such a way that we say "of course!" when we notice that changes in the landscape affect peoples' bodies, and that changes in bodies affect the structure and look of landscapes. As examples like Louisiana's "cancer alley" have shown us, it is no longer a mysterious riddle that the shape of landscapes is part and parcel of the shape of bodies.[6] How can this idea be represented in words? The concept of extensibility is a good place to start. The word *extensible* captures the geographical component—the earthly reality—of what it means for a person to materially exist in places outside a body defined by the edge of the skin. To be extensible is to admit to the spatial reality of which we are all already a part. In this chapter the notion of extensibility is presented as the co-constitution of digestive systems at the Kellogg's sanitarium with the construction of the sewer infrastructure in the city of Battle Creek at the end of the nineteenth century.

THE STOMACH WITHOUT SEWERS

To glimpse into the intestines as they appeared *without* the benefit of a properly functioning sewer system, one can look to the work of pioneering bacteriologist Robert Koch, who in 1884 described the autopsies he performed on victims of cholera who did not live in places with developed sewerage. Initially Koch did not expect to find anything special in the intestines of cholera victims' corpses, but soon he found that "profound and striking intestinal changes" were the norm.[7] Koch began dissecting intestines because he could not find any infectious bacilli in the blood of the bodies, as he had found in his tuberculosis research just two years prior. Focusing on changes in the intestines led him to realize that cholera bacteria came from, and affected most strongly, the digestive system. The content of the intestine was "a bloody purulent stinking fluid," surrounded by a "necrotic" mucous membrane "permeated with superficial hemorrhages," filled with "diphtheria-like deposits, swollen with redness and capillary bleeding." The means of transmission of cholera, typhoid fever, and other fecal-oral diseases was becoming clearer with Koch's bacteriological vision: people were becoming sick because their drinking water was drawn from areas too close to areas of human waste disposal. Reformers subsequently agreed that the solution to reforming the intestines of a growing number of urban dwellers to a healthy state meant the implementation of sewerage.[8] The quality of the concealed organs inside the body, therefore, depended on the shape of the hidden underground of the city.

It is unclear exactly when Battle Creek finished constructing its sewer system, though the earliest applications for sewer taps date from 1892.[9] Provisions from the 1891 city charter outline general guidelines for the

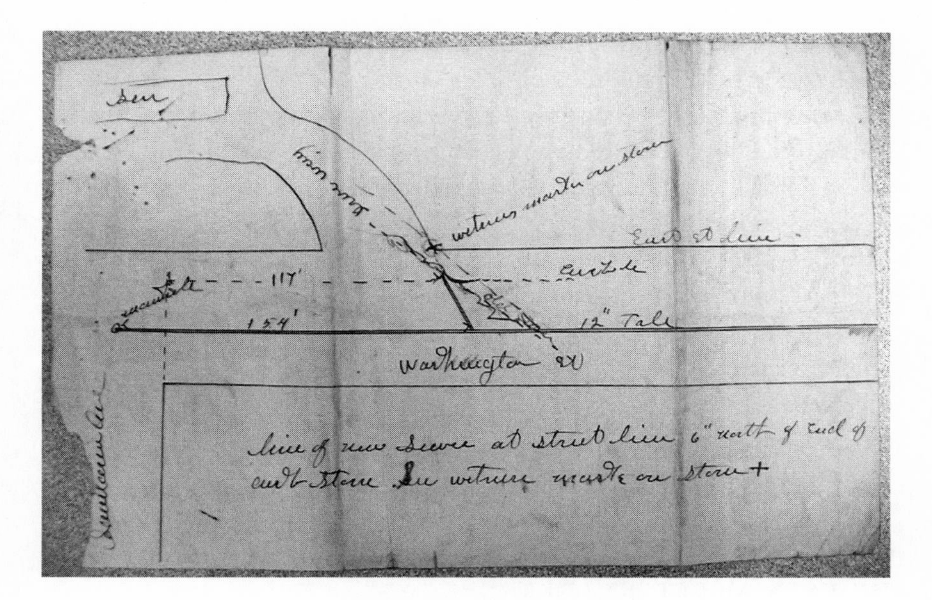

Figure 14. Sketch plan on the reverse side of the application to install a sewer connection at the Battle Creek Sanitarium, 1898. At the intersection of Washington Street and Sanitarium Avenue, the "San" is the rectangle in the upper left-hand corner of the sketch. This map represents the initial phase of extending patients' bodies into the urban landscape of Battle Creek. Image reproduced by the author. Courtesy of the Battle Creek Public Works Department Archive.

development of sewerage, implying that by then contractors had begun to lay the pipes, or were preparing to do so.[10] Assuming that 1892 marks the first availability of a public underground sewer, then there was only a five-year gap between the availability of city-supplied freshwater (1887 in Battle Creek) and its coordinated disposal. By 1892, fresh water was used to transport sewage waste into the Kalamazoo and Battle Creek Rivers, different water bodies than those from which the fresh water was drawn.[11] Battle Creek was not strange in this regard. Due to cost prohibitions, small towns in the Midwest would generally hold off as long as possible before making sewers ubiquitous throughout the town, until the population density was high enough to engorge the landscape with waste.

The "sketch" (figure 14), as it is called in John Kellogg's 1898 sewer tap application, is really a map. Seeing this document as a map is significant because maps indicate spatial ordering on the surface of the earth. To think of bodies as spatially extensible—*constituted by* other objects in the built environment—it is necessary to visualize where and how exactly this constitution takes place in that particular environment. This map is a rescaling of the human body, translating it from the codes of medicine to the codes of engineering, from anatomical drawing to urban plan (compare with figure 15). How, though, is this local street map, which highlights the location of sewer lines and manhole covers, brought into league with an anatomical drawing of the digestive system? And further, how can we read this map *as* a map of digestive systems at the sanitarium? To answer these questions we have to look at discourses concerning disease etiology, waste disposal technologies, and the process of nineteenth-century urbanization.

INDUSTRIAL BATTLE CREEK

For American cities and towns in the nineteenth century, establishing themselves as reputable and progressive was a matter of exhibiting industrial prowess. By 1890 the population of Battle Creek had grown to around fifteen thousand people. At that time it was the fastest-growing city in Michigan, home to farm-machinery, steam-pump, publishing, and flour-milling industries, all in addition to its nascent health food industry.[12] Battle Creek boosters celebrated rapid population growth accompanied by a supposed enhanced quality of life brought on by industrialization. This is an era during which black smoke pouring from factory stacks was drawn onto government letterhead with pride. While on the one hand industry symbolized progress and prosperity, this ethos also encountered opposition. One can, for example, look to the transcendentalism of Emerson and Thoreau for warnings about the dangers of unbridled urban development, a loss of self in the mire of industry and artifice. Another type of reaction to industrialization—one that exhibited a technological approach to the impulse of progress—was the rise of governmental public health programs.[13] As cities grew in population, diseases that would otherwise have occurred within a single farmstead or village began to affect large numbers of people on a fast time scale. Those concerned with public health looked to medical science for answers about the etiology—or causative agent—of intestinal diseases. Debates surrounding the etiology of these types of diseases in the mid and late nineteenth century would influence the design of diet at Kellogg's sanitarium, as well as the design of public sewer infrastructure in Battle Creek. And each of these designs would contribute equally to making and spatially extending bodies in this southern Michigan milieu.

NO MORE PRIVIES

Kellogg's success in making digestive systems that could thoroughly and systematically flush their contents meant that the waste had to go somewhere. Sanitary engineers were sweeping through comparably sized American cities, preaching the perils that would befall these rapidly growing and industrializing cities if their residents did not follow larger, European cities and implement sewage infrastructure. Arnold Clark, a public health expert working at the Michigan State Board of Health, cautioned that "a little village may get along for a time without any special system for the disposal of its waste, but as it becomes more thickly populated, as its soil becomes saturated with wells and vaults, its death-rate naturally begins to increase. . . . A public water-supply and a sewerage system for the disposal of its waste seem to me just as much the business of a city of this size as the lighting of its streets."[14]

He proceeded to rhetorically ask the Battle Creek citizenry what kind of municipality they wanted to construct, for, according to him, there was a direct correlation between the infrastructure of the city's waste management system and the percentage of the population that would die of typhoid fever and other diseases. From urban planning precedents in Europe, it was widely known that without sewers the death rates from what we now categorize as fecal-oral diseases were much higher. Clark made repeated reference to European systems of sanitation, focusing on statistics from Munich and parts of France. European cities served as precedent and proof that diseases were mitigated by the implementation of a sanitary sewer system.

In Battle Creek, the most widespread device used to store human excreta was cement-lined vaults. This dismayed Clark, who called them "relics of barbarism." Even though in 1890 there were about thirteen miles of fresh water mains in Battle Creek, many people still used wells for drinking water, and around five thousand vaults were in use throughout the city for human waste disposal. The proximity of vaults to wells was the problem, as the movement of germs from the vault to the well was believed to be unavoidable over time. Advice from the sanitarium's publication *Good Health* offered similar advice from the same period, though from the perspective of the housekeeper, not of the engineer: "Water from a well or cistern which receives the drainage from a vault, cess-pool or barn-yard, should not be used for drinking or cooking purposes."[15] The problems that could arise from contact with cesspools and waste pits, therefore, were widely known among various parts of society; at issue was implementing good solutions.

The implementation of freshwater and sewage infrastructures in the United States is not a story of co-development. In the 1840s—before sewer pipes were installed in even the largest American cities—pipes carrying fresh water *into* the city coursed through many eastern American metropoles, fed by local streams, lakes, and rivers. Waste produced from this water, including human excrement, was most commonly disposed of in privies, outdoor water closets, or dry earth closets, all of which resulted in collections of waste in the form of the nefarious cesspools. The logic behind this form of disposal was that the land or soil served as a filtration system that cleansed water before it reached the well pump or before it circulated back into the stream, lake, or river. Promoters of the emerging bacteriological vision of health found it difficult to overcome popular belief in the power of the soil to filter out anything bad. As such, strong public support for the construction of sewerage lagged behind the bacteriological evidence that proved its necessity. A common objection to expensive city government sewerage projects was the prevalent belief that the soil was a barrier through which filth could not pass. This led to complacency associated with the vault solution, even as rapid city growth and the use of water closets caused them to overflow and soak the low-lying lands.[16]

MIASMATIC AND BACTERIOLOGICAL THEORIES OF DISEASE

The miasmatic theory of disease was based on an observed correlation between putrefying organic wastes and sick people. Proximity to these wastes, which included items such as food scraps, carcasses, rotting vegetables, and fecal matter, increased the likelihood that someone would acquire a disease. The bad-smelling air associated with decaying organic matter was considered the disease vector, meaning that individuals were affected directly by the source of putrefaction rather than by each other. In other words, the miasmatic theory of disease was not contagionist—one could not become ill by being close to someone else with a particular illness; instead, one had to be close to the source of foul smells.[17] Integral to the miasmatic theory of disease, therefore, was an awareness of one's surroundings. Dumping organic waste near a house would create foul odors, and could bring sickness to all living close by. Ella Kellogg reflects this stance in her instructions for housekeeping: "The ventilation of the kitchen should be so ample as to thoroughly remove all gases and odors, which . . . render to some degree unhealthful every other portion of the house . . . the ventilation of the kitchen ought to be even more carefully attended to than that of a sleeping room."[18]

Ella Kellogg's near paranoia about ventilation is a direct descendent of the miasmatic theory of disease, an etiology in which the surrounding environment plays a crucial role in the maintenance of health. Even if for the "wrong" reasons, Ella's promotion of miasmatic etiology succeeded in upholding powerful links between environment (the indoor environment, in her case) and health at the sanitarium.

For most people in rural and semi-rural America, miasmas were the most popular way of understanding diseases well into the late nineteenth century. But beginning in the 1870s bacteriology emerged from the corners of chemistry laboratories as a new way of understanding disease, and it quickly began to spread to the educated and governing populations. Bacteriology introduced microscopic germs, which replaced foul air as the disease vector, as the way to imagine what disease was, where it was, and how it spread.[19] Because miasmas were so ingrained in the popular imagination of how disease worked, though, convincing people that germ theory was a more accurate way of conceptualizing and combating public health problems evolved into nothing short of a duel between germ theory scientists and budget-conscious citizens, who were concerned about the cost of building enormous sewer systems. The debate about the cause of gastroenterological diseases had major implications for how cities were built.

Controlling what kind of matter, germs, nutrients, filth, food, or bacteria entered into the digestive tract, as well as the style in which they were purged from the digestive tract, defined health for John Kellogg. But this process also increasingly defined health for Battle Creek's public sanitary engineers. For Kellogg, the digestive system was the site of physical and moral betterment, the body organ with the highest potential to make the most change in the health of an individual. Like an alchemist, Kellogg used food as the raw material reactant that would transform the stomach, intestines, and colon into efficiently operating precious metals that radiated wellness to the entire body. For the sanitary engineers, the digestive system was the site of defense against, and the staging ground for, serious disease. It was the body organ responsible for incorporating and harboring deadly microorganisms, and for dispersing them wantonly in privies and outhouse closets, putting fellow city dwellers at risk. For sanitation workers, public health would be remade by what was *not* introduced into the digestive system—that is, fecal germs that found their way into the fresh food and water supplies. Sanitation workers, and also John Kellogg, each used the digestive system as a fulcrum of health, whether by adding the right materials or taking the wrong materials away.

At the heart of the issue of urban landscape reform was the transition from the miasmatic—also known as environmental, or filth-based—theory of disease to a bacteriological, or germ-based, theory of disease. Discourse in medical science was undergoing an epistemological sea change during the time that Kellogg formulated his diet prescriptions, and—not coincidentally—during the time that Battle Creek grew into its heyday as a place of industrial activity.[20] With Battle Creek's increasing industry and the attendant growth in population, public health officials at the local level became very interested in explaining the roots of diseases in such a way that health threats could be solved. With more people came more human waste, and the fears of rampant communicable diseases grew. One official from Lansing named Erwin Smith, for example, commented on the crowded living conditions found in southern Michigan towns, and the suggested methods of waste disposal: "The soil should at all times be kept perfectly clean, and on no account should excreta of any sort be thrown into the door-yard, or stored in any vault or pit, to ferment and poison the air and the soil. Every city and village of any size should endeavor to secure sewerage and a proper water-supply."[21]

Smith was adamant that typhoid fever, and cholera in particular, were traceable to filthy, foul-smelling soil, and he was trying to convince people to spatially separate drinking water wells from waste pits. But he went on to talk about the introduction of *germs* into the vaults, demonstrating that in this sea change, like any sea change, the transition between miasmas and bacteriology was not sharp. Each of these disease etiologies, and the practices they spawned, blended together in the closing decades of the nineteenth century, as might the back currents and mixing eddies of a new tide. Hence, for Kellogg's practice of medicine at the sanitarium, as well as for the city's sanitary engineers, actions taken to cure intestinal health were grounded in both miasmatic and bacteriological theories of disease, and often at the same time. With this background in mind, we will be able to see more clearly how the practice of digestion at the sanitarium was tied with the building of the city's underground sewer infrastructure, and how materially constructing a new city in fact materially constructed new bodies as well.

With the rise of bacteriology, public health workers pleaded with city governments to install sewer systems that would put a comfortable, albeit imprecise, distance between people and their collective organic wastes. This reasoning should have worked equally well using the logic of miasmas since spatial proximity to sinks of animal and human waste increased contact with bad air. Why not remove the waste out of sight and smell? Public health reformers around the 1880s tended to hedge their arguments, using the languages of foul air as well as germs to promote the installation of urban sewerage. No matter how they got there, though, consensus was building around a conclusion that would demand the installation of sewerage, an extension of the process of digestion concealed underneath the grade of the city's streets. In figure 14 we peer into this hidden world of main lines, manholes, curb lines, and tiles to glimpse the engineer's dream of protecting the digestive system from itself by creating a new piece of the human biological ensemble.

KELLOGG AT THE CENTER OF IT ALL

While John Kellogg was refiguring the movement of food through the digestive systems of his patients, teams of engineers were refiguring the movement of sewer water through the subgrade of Battle Creek. Each was trying to apply the clear, convincing, rational language of science—in Kellogg's case, to the body, and in the engineers' case, to the city. Scientific eating and scientific sanitation were coming together, and each was drawing from the same logic of microscopic, chemical analysis to shape the material world in a way they thought would benefit society. And surprisingly, each of these fields—in late nineteenth-century Battle Creek—was led by John Kellogg. In 1879, as a member of the Michigan State Board of Health's Committee on Disposal of Excreta and Decomposing Organic Matter, he wrote a report outlining the benefits of sanitary science at the municipal scale. In it Kellogg addressed with great urgency the need for Michigan towns to implement a policy that would ensure the widespread use of sanitation technology. The concern was that human excreta, and the "fetid, pungent odors which are poured upon the air by such a hotbed of disease," carried and spread illnesses through the towns. Though water-carriage sewer systems were unrealized in Michigan towns in the 1870s and 1880s, this fact did not stop Kellogg from promoting them, nor from decrying the "evils" of privy systems, the most "abominable nuisances" in which feces piled up and decayed, threatening the well-being of the entire community.[22] Promoting annual household inspections from sanitary scientists for all residents, Kellogg clearly brought an awareness of external conditions to his practice of medicine. Until 1891 Kellogg recurrently chaired the Committee on Disposal of Excreta, bringing to it his medical expertise on the biological processes of digestion. His unique position as a gastroenterologist trained in medical science, coupled with his role as director of a small society with pressing environmental needs, made him better positioned than anyone to perceive digestive systems as part of the urban infrastructure. Kellogg more than anyone, that is, could have imagined the longed-for sewer system as an elegant extension of the intestines through (and out of) the city. The scientific eye that saw harmful bacteria festering in the intestinal canal was the same eye that saw disease germs festering in the cesspools around town, threatening to poison everyone.

Theories of health and disease were influencing how public urban infrastructure was being constructed. In another one of his publications as a member of Michigan's State Board of Health, Kellogg captures the blending of etiological theories that were so prevalent, with which doctors as well as city engineers grappled. Speaking to the problem of decaying organic matter, he wrote that "Chemistry brings to light poisonous gases, the presence of which is confirmed by the sense of smell; but the microscope makes a still more important discovery, viz.: the presence of myriads of minute specks of life, to which the name of germs has been attached."[23]

In this passage from 1882 Kellogg portrays the underlying challenge that he and other reformers faced in trying to convince taxpayers to support new urban waste disposal infrastructure. Allaying the financial concerns of city governments that were pressured to construct new sewer systems began with reports from medical science. For Kellogg to convince people of the need to construct water-carriage sewer infrastructure in Battle Creek in 1882, he had to scientifically demonstrate how disease worked in both the miasmatic and bacteriological etiologies.

In this period of sea change, Kellogg knew well that decaying organic waste threatened the very health of the bodies he was building at the sanitarium, and therefore sought to remove it swiftly, cleanly, and efficiently using any epidemiological rhetoric necessary. His focus on the removal of fecal matter from the location of the sanitarium indicates the spatiality of the entire health practice. Kellogg was aware of the necessity to move human waste. His efforts to reshape the whole of Battle Creek in a way that would create the right spatial relationship between the digestive systems of his patients and their decaying, organic evacuations is where Kellogg's geographical sensibility can be found.

KELLOGG CURES AUTO-INTOXICATION

How can we bring together John Kellogg's religious-hygienic philosophy with the science of digestion and the technics of urban engineering? The concept of "self-poisoning," the technical term for which in the 1890s was "auto-intoxication," is a helpful place to start. Introduced by the French pathologist Charles Bouchard, the notion of auto-intoxication appeared after an era characterized by the discovery of bacteria, or, as they were more generally called, "microbes." As microbes were increasingly used to explain disease and contagion, Bouchard researched and wrote extensively about the relationship between microbes and bodies.[24] He became fascinated with how a body could live with such perilous beings but avoid illness. He wanted to know how the body dealt with dangerous bacteria, and how one could become poisoned from within. This line of inquiry led Bouchard to imagine the human body as always on the precipice of disaster, sickness, and death. If people were constantly surrounded by, consuming, and actually composed of potentially harmful bacteria, then responsibility for health became a moral responsibility of the individual, who needed to be on guard and take the proper actions at all times in order to stave off the onslaught of poisonous microbes. It was a tightrope walk—living with and expelling bacteria—that the masses could not have possibly been equipped to handle without education.

If it was solely for the benefit of his sanitarium business to adopt Bouchard's philosophy, Kellogg never admitted it. Instead, Kellogg seemed to adopt auto-intoxication with earnest intent and purpose. By integrating it into his hygienic program, Kellogg taught his patients how to think about what was going on in their intestines, how to conceptualize problems of digestion, and how to solve them.[25] The basis of auto-intoxication comes from the stagnation of food in the digestive system, as the quality and/or quantity of poisonous microbes could become harmful enough to cause disease. Kellogg practiced what Bouchard had written about auto-intoxication. In private consultations, patient appointments, public lectures, and his own publications, Kellogg transformed ideas that grew from Bouchard's laboratory into a lived experience for his patients. It was Bouchard who broke down the barrier between infection and contagion, using germs to explain both at the same time. Infections were nothing more than germs that originated in a household privy, while contagions were the same germs as found passing through a living organism.[26]

While Bouchard promoted enema cleansing for relieving symptoms of diseases brought on by auto-intoxication, it was Kellogg who developed enemas into a systematic practice, and who gave sparkling clean intestines the curative weight of pharmaceutics. Because of Bouchard, Kellogg took stagnation seriously, and so—with his characteristic drama and flair—developed a system of healing that sought to evacuate bacteria from inside the body. Kellogg thought the body was a "factory of poisons," stating that "the amount of poisons produced in the body and the extremely poisonous nature of these substances may be inferred from the rapidity with which death occurs when there is any serious interruption in the process of poison elimination."[27] One of the pillars of practicing medicine at the sanitarium, then, was creating an environment in which there would be no interruptions in the activities that made "poison elimination" possible. The place of the sanitarium was paramount to making this happen. It was a destination at which the performance of Kellogg's hygienic philosophy could unfold without the pestering duties of daily, urban life. The place made the medical practice real, re-*placing* people's bodies from their homes into the milieu of southern Michigan, while replacing their routines with new ones. During the span of their stay at the sanitarium patients left the cycles and patterns of wherever they came from and entered a retreat-like space that included scheduled exercises, hydrotherapies, treatments, consultations, and lectures. They were living inside the social and architectural manifestations of John Kellogg's philosophy, actors in the live performance of his dream. And this dream of perfect health through perfect digestion could not have progressed as it did without tapping into the twelve-inch main sewer line that ran down the center of Washington Street in front of the sanitarium.

One of the major recurrent themes of this temporary new reality that patients experienced at the sanitarium was a focus on food: what to eat, when to eat, how much, and how often. For Kellogg, the sanitarium kitchens were nothing less than pharmacies, serving custom-designed and controlled meals that would reduce the amount of bacteria introduced into the digestive system, and therefore the possibility of auto-intoxication. Kellogg wrote that "nearly all chronic ailments, especially the disorders of the heart, arteries, and kidneys, are the result of the absorption of poisons from the intestine, poisons which are for the most part produced in the lower part of the small intestine and in the colon through the action of germs upon undigested food remnants."[28]

Stagnation of indigestible food was the problem. Food could become trapped at various muscular valves—termed "gates" by Kellogg (see figure 15)—inside the digestive system, stay there, putrefy, and eventually lead to chronic illness in other parts of the body as far away as the heart or back, and could even affect a person's mental state and disposition. Consistent, clean, free-flowing circulation of food through the intestines was the remedy, and could not be achieved without a "hygienic" diet made up of bran, fruit, and vegetable variants, nor without a defecation schedule that approached a part-time job.

The sanitarium philosophy was that "a colon filled with putrefying residues and wastes is a constant menace to the body" and "a potent means of inducing premature senility."[29] As one former patient put it, "the Doctor's great slogan was 'Keep the Colon Clean.'"[30] The idea for Kellogg was to withhold foods that produce putrefying residues so as to maximize the efficiency with which the intestines could transform food into nutrients, and to minimize the burden placed on the "poison-eliminating organs"—the liver and kidneys. Food that was prescribed was often done so for the ease and speed at which it could move through the digestive system. One of the prominent ways to achieve this efficiency was by eliminating meat from the diet, using the logic that "anything that will putrefy outside of the body will putrefy in the colon." Patients were instructed to evacuate at least three times per day, with an eight- to twelve-hour lag time between eating and defecating. In a time-lapsed mapping of this schedule, the sanitarium Diet Service Department drew a series of diagrams of the digestive system (figure 16). Kellogg and his clinical staff marked successful healing by a patient's ability to train his or her body to match this type of evacuation schedule. If the schedule was broken, a patient would be scrutinized for what he might have done to derail the order imposed by the hygienic philosophy. Bad timing meant that residues were stuck at one of the intestinal "gates," putrefying into a mass of bacteria that would spread to other organs.[31] As we saw in chapter 2, one common solution was a reformulation of the patient's diet in a way that increased the amount of

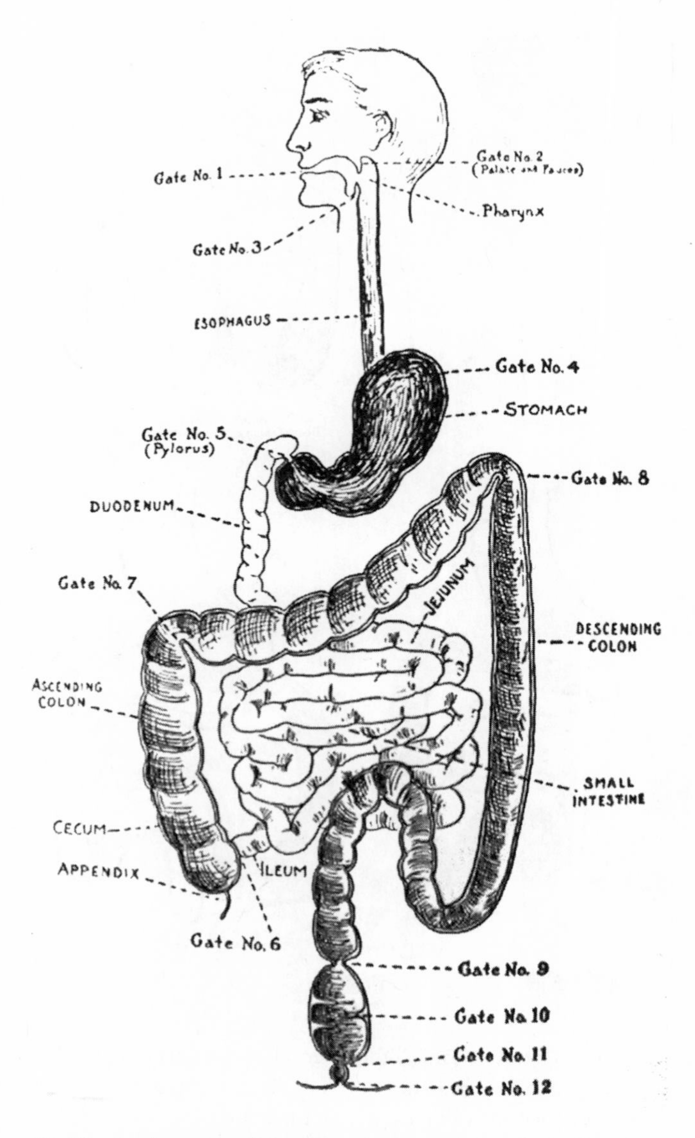

Figure 15. The twelve gates of the alimentary canal. In John Kellogg's words, "many of the most serious disorders of digestion are the result of disturbances which occur at the food gates." Here the body is reduced to the digestive system with a head attached, highlighting the great significance of these organs that Kellogg attributed to healthfulness. Image reproduced from John H. Kellogg, *Constipation: How to Fight It* (Battle Creek, MI: Good Health Publishing Co., 1913). Courtesy of the Bentley Historical Library, University of Michigan.

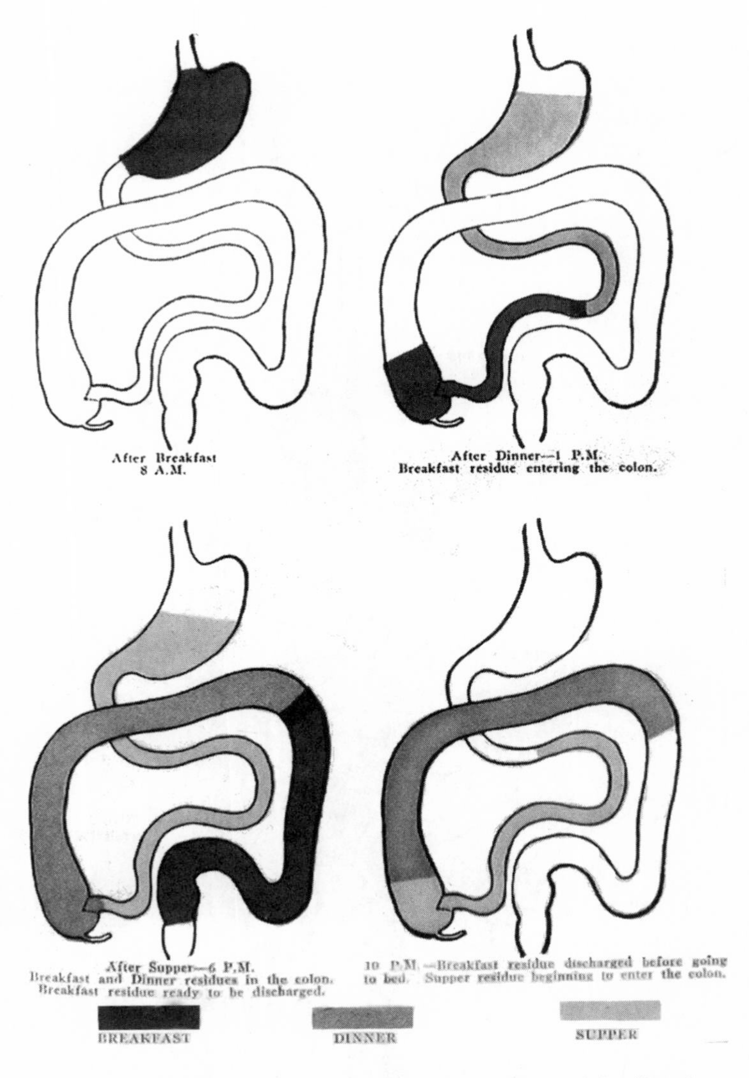

Figure 16. This "Digestive Time Table" maps the contents of the digestive system in an ideal twenty-four-hour period. The drawings correspond to three meals eaten within a single day, and their subsequent movement through the digestive tract in Kellogg's ideal order and timeliness. This anatomical diagram excludes what happens to the digested food after it has been evacuated, leaving out an important part of the body's infrastructure. The sewer sketch in figure 14 answers this question of where the digested food goes, and is a prominent example of corporeal extensibility. Image reproduced from *Diet Helps for Doctors,* written by the Diet Service Department of the Battle Creek Food Co. (Battle Creek, Mich., 1928). Courtesy of the Bentley Historical Library, University of Michigan.

fibrous foods taken in. In other cases a series of enemas would be administered, further outsourcing the process of digestion away from the body, to a piece of technology called the colonic machine.

THE COLONIC MACHINE: OUTSOURCING EVACUATION

The colonic machine (figure 17) was a technological implement that did the work of the intestines and colon when they could not keep up with Kellogg's prescribed defecation schedule. This glass bowl object with its set of valves is a biotechnology that materially extended digestive systems in a way particular to Battle Creek's landscapes of fresh- and wastewater infrastructures. The machine worked by introducing three pints of cool water into the colon with a long rectal tube, after which patients were instructed to hold in the water for ten minutes.[32] The fresh water used for this step arrived from Goguac Lake, a source from which Battle Creek drew all of its fresh water at that time (see figure 18).[33] A sanitarium publication reported that its annual consumption of water was ten million gallons.[34] Part of this figure included the water used for hydrotherapeutic procedures, including the colonic machine. Bringing the water from the lake to the city was a matter of drawing it out with a steam engine pump, then letting it fall by gravity for two hundred feet over a distance of two miles to the northeast. In 1887 a Chicago-based engineering firm completed the infrastructure that brought this fresh water to the city of Battle Creek, and, by matter of course, into the colonic machine.[35] As the fresh water from Goguac Lake was introduced into the intestinal tract, it mixed with the partially digested food and bacteria that were feared to be stuck at one of the lower intestinal gates, cleansing and releasing residues that could bring on auto-intoxication. The soiled water was then released back into the machine, through which (starting in 1898) it drained away into the sewer system.

By connecting patients to the colonic machine, every step in the biological process of digesting and excreting food was replaced by technological contrivances. Fresh water came in through one pipe, while bacteria-laden water exited through another pipe (see figure 19). The fresh water that entered through one pipe came from the steam pump at Goguac

Figure 17. One of the colonic machines used at the sanitarium. Its purpose was "to treat toxicity and constipation" in patients. This device was the object that immediately extended digestive systems to the freshwater and wastewater infrastructures of Battle Creek, making lakes, rivers, and pipes all part of the process of digestion at this time and place. The device stands approximately 0.75 meters high. Photo by the author, taken at the Seventh-day Adventist Heritage Center, Battle Creek, Mich.)

Figure 18. In the 1890s Goguac Lake was clearly in the rural hinterland of Battle Creek. This lake was the source of all the sanitarium's fresh water. John Kellogg had access to a lakeside resort and pavilion here that was used by patients and guests. Image reproduced from *Illustrated Atlas and Directory of Free Holders of Calhoun County, Michigan* (Fort Wayne, IN: Atlas Publishing Company, 1894). Courtesy of the Michigan State University Map Library.

Lake. After it cleansed the intestines, it exited through a series of sewer pipes below Washington Street, on to the Kalamazoo River, from whence it flowed westward to Lake Michigan. The colonic machine was the piece that brought the body into this movement of water and feces. As it relieved the colon from the task of purging its own waste, it spatially relocated the process of digestion to the built environment of the urban landscape. The entire system was doing the work of the digestive system—that is, safely removing potentially poisonous food waste from the body—without the digestive system having to fully complete its function. With this combination of biotechnologies a patient's digestive system was outsourced and spatially extended into the southern Michigan landscape, completely isolated from the germs that could infect it, while completely reliant on the steam pump from Goguac Lake and the sewer tap that was installed at the sanitarium in 1898. The spatially extended body turned out to be a healthier one in this context. The object-organism cyborg made up of guts, colonic machines, and water-carriage infrastructure was a safer way to live, as the possibilities of auto-intoxication and fecal-oral infection were greatly reduced by making people exist as part of the city.

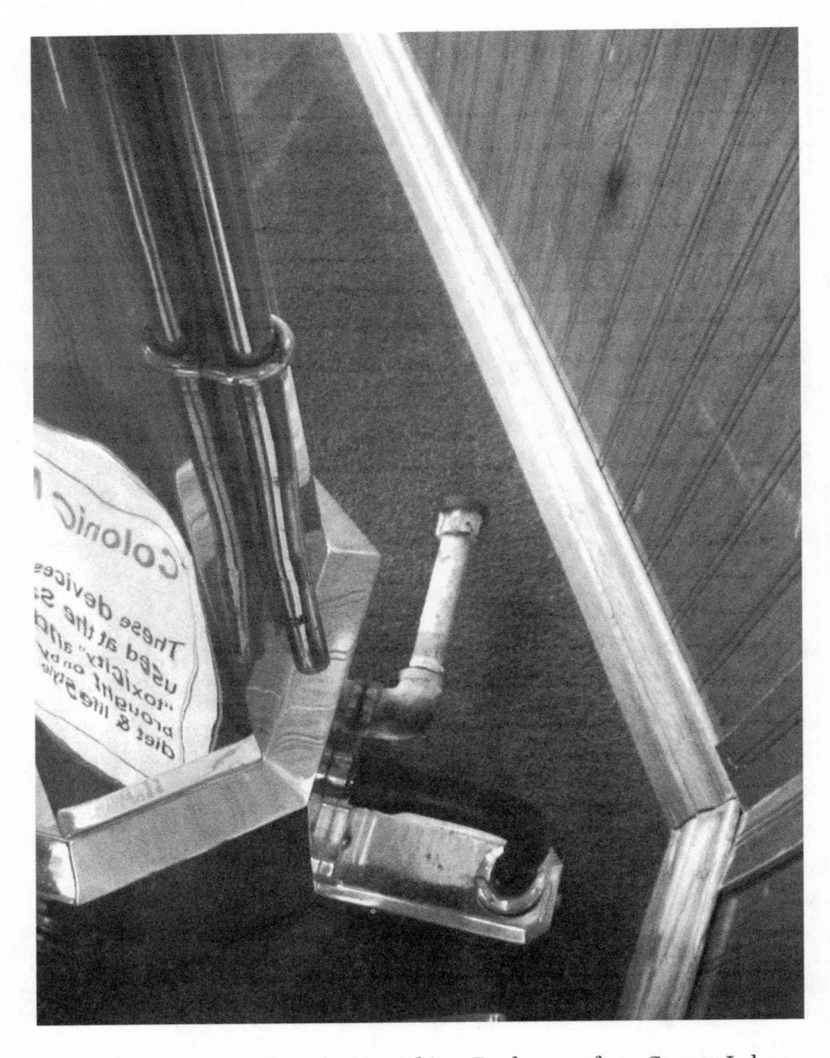

Figure 19. Close-up of the colonic machine. Fresh water from Goguac Lake entered through the white pipe into the colon, and the used water exited through the black pipe into the sewer system. This machine outsourced the work of the digestive system, materially extending it along both freshwater and wastewater infrastructures. Photo by the author, taken at the Seventh-day Adventist Heritage Center, Battle Creek, Mich.

CONCLUSION

In Kellogg's taxonomy of the body's "food laboratories," he compared the mouth to a mill, the liver to a refinery, and the colon to a waste disposal system.[36] His metaphorical bridging between the digestive system and the urban manufacturing landscape was direct. Such comparisons harked back to the significant urban infrastructure changes occurring in late nineteenth-century America. Cities were developing new sewerages, underground hidden pipes not unlike the piping of the hidden intestines inside the body. As John Kellogg scrutinized the innards of the body, he also scrutinized the innards of the city, giving them both heightened attention as a matter of public health in order to treat diseases. The flushing of waste and the movement of water through the city was a macrocosm for the movement of food, water, and waste through the body. In each case, stagnation was assumed to be deleterious. Food had its curative benefits, but if it stopped moving it had the power to harm, in the guts, in Battle Creek's cesspools, and in the pipes of the city's sewerage.[37] To see the movement of human waste as active and tubular in both of these settings is to wonder about the intellectual origins that recognized and promoted this process that sends poisonous matter somewhere else, to a location far from bodies where it can no longer act as an auto-intoxicant. The continuation of the body's intestinal tubing into urban sewerage is an extension both conceptually and materially. Sanitary engineers spatially extended the digestive system into the landscape by building a body-world that reached beyond the confines of the skin. Connecting the circulation of food through the tubes of the body and the circulation of discharged food through the tubes of the city is a matter of bringing these two geographic scales into one view such that the infrastructure of the city becomes an extension of the body, and the body becomes part of the landscape.

5 The Systematization of Agriculture

This chapter examines two intimately connected practices in the 1890s that were becoming enveloped into the analytical gaze of science in the United States: growing food and eating food. The application of chemistry and mechanization to fields, and the rise of private seed warehouses built an agricultural science that affected the availability of foods utilized by John Kellogg. The knowledge produced by government-funded agricultural science translated into the gastro-based medical practice at the Battle Creek Sanitarium. On the one hand, there is an obvious categorical connection between growing food and eating food. On the other hand, though, this translation from science on the farm to science at the table was not obvious when Kellogg was at the pinnacle of his health practice. The material outcomes of that agricultural science changed the way food was consumed and changed the quality of the digestive systems for patients in Battle Creek. In this chapter, traveling farthest on the transect leading away from the location of the sanitarium, the agricultural landscapes in southern Michigan and beyond complete the material circulation that made the geography of digestion in Kellogg's universe.

Two seemingly disparate spheres of society—the pharmacological consumption of food, and the positivistic production of crops on American farmlands—were undergirded by the same scientific epistemology that looked to chemistry for a way to maximize efficiency and production, both for the soil and for bodies. With the application of this chemical knowledge—more potent fertilizers, for example—came the production of food in higher quantities. And as we have seen in chapter 2, this same reliance on chemistry was the foundation for how Kellogg diagnosed and treated a range of health disorders via the digestive organs. While it might seem as though the apparatus of chemistry was applied in two entirely different places—the land and the body—with two entirely different aims, the congruencies they share remind us that the cultural and linguistic veils we use to divide making food from eating food are never as strong as they seem. If we look carefully enough (and sometimes if we simply look at all), we can always find one in the other; the movement from food production to food consumption is not linear, but cyclical. In some places the confluence of these two spheres is more immediate or obvious. In Kellogg's practice of health at the turn of the century, the point where agricultural science and nutritional science met was clearly the stomach. Kellogg wrote that "the general tendency of thought in any nation may be determined by the character of the national diet. True as this principle is when applied to the body in general, it is especially true in reference to the stomach."[1]

In this 1896 statement, he posits that a "national diet" makes bodies and tendencies of thought in a particular and shared way. But the making of the stomach is the means by which a national character is ultimately forged. Bringing this into the geography of digestion, then, I interpret the stomach as connected to a wider landscape of food production, further actualizing the theoretical concept of geographical extensibility among objects. Creating a national stomach could not have been imagined without first creating a national agricultural landscape that could provide the necessary foods.

Midwestern agriculture in the late nineteenth century was becoming increasingly mechanized, reliant on science, and large in scale, able to produce en masse the grains that would cure the ailing modern stomach as prescribed at the sanitarium.[2] "Something is wrong with the modern stomach," wrote Kellogg, "the average stomach in civilized lands grows weaker year by year, and disorders growing out of indigestion multiply with alarming rapidity."[3] As discussed in chapter 2, the primary solution for taming the wilds of a premodern stomach was a diet that rested on the availability of products designed with whole grains, fruits, and nuts, meant to ease the work required of the digestive system and to shorten the amount of time food would spend inside the intestinal canal. Understanding how foods were made that would interact with the body in this way completes the geography of digestion, connecting the work of the intestines to landscapes of agricultural production. Where did all these foods come from, and how were they grown? Just as gastrointestinal tracts at the sanitarium were becoming quantified and homogenized with the language and tools of chemistry, a similar discourse of rationalization was taking place on the mighty grain fields and fruit farms of southern Michigan.[4]

Chapter 4 described how food became part of an assemblage that included lake, steam pump, cesspool, intestine, sewer, and river. As we move away from the body and the digestive system we encounter various machines and infrastructures in places along the way. Now, as we continue to move on a transect through space away from the body, digestion enters a new type of assemblage, with new connections to the material world around it. Writing about the process of metabolism—the movement of materials as they course around the earth—geographer Erik Swyngedouw points out that "metabolism . . . simultaneously implies circulation, exchange, and the transformation of material elements. When matter moves, it becomes 'enrolled' in associational networks that generate qualitative changes and qualitatively new assemblages."[5]

Shifting our view outward, away from the city, takes us to the origin of the food material that Kellogg employed so scientifically at the sanitarium. As we look at food in the fields, as opposed to the plates of patients, we are led to a qualitatively new assemblage. Describing it relies on exploring mechanical and biological technologies that made the sanitarium's food available.

MIDWESTERN AGRICULTURE IN THE SECOND HALF OF THE NINETEENTH CENTURY

The deliberate use of food to reform bodies at the sanitarium came out of a deliberativeness that agricultural scientists were using to reform the shape of rural landscapes. Here I describe the economic state of agriculture and the "discontent of the farmer" in this period, then move on to two technologies—threshing machinery and mail-order seed banks—that affected the way food was grown, extending the geography of digestion to the hinterlands.

Researchers at the U.S. agricultural experiment stations sought to address what was becoming widespread rural economic distress. By the late nineteenth century it was widely acknowledged among U.S. farmers that they needed one another to overcome their economic difficulties. A Michigan farmer in 1892 made the following lament:

> The all-absorbing topic with the farmer at this time is, what shall we plant or sow that will bring us some money this fall, for there is the hired help to pay and the inevitable taxes, which like the poor, 'are always with us,' and nothing but the cold hard cash will answer. Cattle are cheap, we cannot depend on beef or butter for a profit; our clover did not catch last season, and cold weather and bare ground have caused what did survive to heave out of the ground, and it is ruined; so we cannot depend on any surplus hay to sell, and of course no clover seed to sell and help out.[6]

The dying independent, autonomous, yeoman farmer caused much concern, and encouraged a number of political responses—such as the Farmer's Alliance and the Grange—which contributed to a larger populist movement that brought farmers together with one another. The attitude of these organizations was that the solution to farmers' market troubles rested with the farmers themselves; they were on the whole extremely suspicious of government advice. In 1896 Harvey Wiley, then chief chemist of the U.S. Department of Agriculture (USDA), attempted to assuage the concerns of farmers by saying that ground-up organization alone could not save farmers. Rather, according to Wiley, they needed guidance from outside their circle to help increase the quality and quantity of yields.[7] In short, he thought that they needed new forms of knowledge, and that this new authoritative knowledge would come from agricultural scientists. In the view of the USDA, agricultural science would offer farmers an alternative to both political organizing and traditional farming practices, the results of which had not been effective enough in overcoming a rapid shift in agricultural economics toward food-for-market production. In essence it offered a "third way" for farmers to adapt to a dramatically and quickly changing agricultural economy.

Proponents of scientific agriculture claimed that in the past it was only the abundance of cheap, fertile, accessible land that saved farmers from their lack of knowledge about farming. According to the scientists, methods of irresponsible farming, defined by unnecessary labor and waste, emerged from this ease of spatial mobility. When the soil in fields became "worn out," farmers were able to move to a new piece of land—generally in a westward direction over the decades—relatively easily. A confluence of population intensification and competitive agricultural markets in the late nineteenth century created a situation where something had to change. Responding to what they saw as farming's slow development in an era in which other sectors of society were adopting scientific principles, university and governmental agricultural scientists were eager to rationalize practices on the farm.

This rationalization took the form of what was called *systematization*. Creation of a farm system included the following actions: diversify and rotate crops, use manure and artificial fertilizer, organize a farm layout and use barn architecture that is conducive to efficiency, and establish a strict pattern of activity for chores and housework. The idea of systematization promoted by government officials defined farms as "a group of discrete yet independent and hierarchically organized parts," or in other words, like a factory.[8] The notion of a division of labor on farms, along with efficient spatial organization, took a page from the book of the industry-town model. For engineers, farms and factories were likened for two reasons. First, farms transformed materials into consumable goods. Second, farmers in the late nineteenth century were becoming subjected to a new division of labor that took away their power.[9] To industrialize farms meant to take farmers away from the quotidian decisions that went into land management. Instead, this responsibility fell to external contractors, many of whom were trained in business rather than agronomy.

FOOD CHEMISTRY IN ACTION

Kellogg was acutely aware of the link between food chemistry and physiology, recognizing that "the function of a tissue or organ depends upon its structure. The structure of every cell and fiber of the body depends upon the quality and quantity of the material absorbed from the alimentary canal."[10] Relating food's material quality to the structure and function of cells in particular organs is another indication that Kellogg was deliberate in his prescriptions. For Kellogg, foods were precision tools, affecting the digestive system in ways that would then affect the rest of the body. His list of prescription diet tables were "based upon observations made in connection with the exact methods of analysis of stomach fluids and the study of digestive disorders."[11] Under this approach, the work of sanitarium chemists intersected with the work of staff in the experiment kitchen. The contents of patients' stomachs were brought into the chemical laboratory for analysis, and the results dictated which one of at least twenty-five diets should be administered. Analysis measured mostly for hydrochloric acid and chlorine (figure 20). The health of the patient represented in figure 20—who was "thin in flesh and somewhat anaemic [*sic*]"—was reduced to a quantitative measure of stomach acid. In this case Kellogg claimed that the patient's stomach produced a lot of acid, but not enough of it became useful for the process of digestion because the required chlorine was consumed by combining with "irritating" foods, such as pickles, radishes, onions, and sauces. The resulting was that "the digestive product [gastric juice] is vitiated in character, so that the work of the stomach is almost valueless."[12] The diet cure for hyperpepsia included foods that reduced the overall chlorine output, like rice, soft-boiled eggs, stewed fruits, and puréed peas. Chemical results such as these dictated how food was prepared in the sanitarium kitchen, where cooks utilized quantitative guidelines to construct the perfectly measured medicinal meal. These processes were deliberate, the actions triggered by the chemical-numerical results of the contents of the stomach. This demonstrates specifically one of the ways that Kellogg overlaid a quantitative gastronomy over the body, one that would "thicken" the flesh and cure anemia.

Kellogg's deliberate use of food as medicine—as in this example about anemia—emerged directly from government-funded *agricultural* research, which sought to rationalize the practice of growing food. At first it may

Total'acidity, (A)	350 gms.	(A') 366 gms.	(.180–.200 gms.)
Coefficient, (*a*)	.73	(.86)	
Total chlorine, (T)	.496	gms.	(.300–.340 ")
Free HCl, (H)	.256 gms.		(.025–.050 ")
Combined chlorine. (C)	128 "		(.155–.180 ")
Fixed chlorides, (F)	.112	"	(0.109 gms.)

Figure 20. These results from a stomach analysis at the sanitarium laboratory indicate a case of "hyperpepsia." Image reproduced from John H. Kellogg, "The Treatment of Hyperpepsia," *Modern Medicine and Bacteriological Review* 3, no. 4 (1894): 94. Courtesy of the Bentley Historical Library, University of Michigan.

seem odd that research in growing plants and raising animals would translate so directly to human nutrition, but it was the shared language of efficiency and pinpoint control that would put them in concert. Human nutrition emerged as a respected scientific endeavor in the United States in the 1890s, closely tied with the growth of research in scientific agriculture.[13] To clarify the intellectual environment in which Kellogg developed his quantitative gastro-based medicine, it is important to detail (1) the emergence of scientific agriculture in Germany, (2) how that knowledge migrated to the United States, and (3) how it then became applied to human bodies. Doing so will show us that food in this story was subject to the same scientific gaze in at least two different points during its path from seed to stomach: once in the fields, and once when it entered the body. In addition to showing the material connections between the food we grow and the resulting bodies we create, this history of science shows us how, Kellogg used systematization—key components of which are quantification and homogenization—to make and sell the promise of a deliberately, rationally produced body at the Battle Creek Sanitarium.

SCIENTIFIC AGRICULTURE IN GERMANY

German institutions responsible for training agricultural scientists have had far-reaching impacts. For example, from them emerged the German schools of forestry, where the famous American conservationist Gifford Pinchot was trained. The eventual outcome of the German agricultural schools was a guiding idea of maximizing yields—whether from the farm or the forest—for profit.[14] In Pinchot's case, this approach led to the damming of the Hetch Hetchy Valley in Yosemite National Park for the utilitarian capture of fresh water for the San Francisco Bay Area, a landmark case study in the American conservation movement. Pinchot, however, is just one instance of the many American scientists who made multiyear pilgrimages to Germany in order to advance their skills in the science of farming, so that they could return to the United States and disseminate procedures on how to more efficiently grow crops.[15] During the second half of the nineteenth century it was easier for European researchers in chemistry to receive funding for "theoretical" pursuits than it was for American researchers, who at the time tended to be funded for research that resulted in "practical" advice to farmers.[16] The need of American scientists to train in Germany belied a philosophical debate over the contents of agricultural education and training between the two continents. In 1840, for example, French political writer Alexis de Tocqueville "lamented a lack of emphasis on theoretical science in the United States."[17]

In 1813 the Englishman Humphry Davy wrote one of the first theoretical agricultural chemistry books, titled *Elements of Agricultural Chemistry*.[18] The major dilemma he addressed in this work was the fact that humus—decomposed organic matter near the soil surface—was apparently insoluble, and yet was found in plants. The burning question of the day was how the humus became dispersed throughout the plant. Davy was very concerned with the processes of circulation and incorporation of humus, topics that framed debates in agricultural chemistry for the rest of the century. Historian W. A. Shenstone describes Davy's humus theory in light of Davy's prominent successor, Justus von Liebig, who would become the godfather of agricultural science in Germany:

> It had come to be supposed, prior to the date of Liebig's earliest writings on agricultural chemistry, by many chemists and agriculturists, that this vegetable mould was the source of the fertility of these soils. By an extension of this idea many vegetable physiologists ascribed the fertility of all soils to its presence, and even regarded it as the chief nutriment of the plants; it was supposed by them that the humus was extracted from the soil by the roots of the growing plants.[19]

In 1840 Liebig deflated Davy's humus theory by showing that humus did not contain all the elements necessary for plant nutrition. Liebig introduced a new theory—the nitrogen cycle—that was still based on circulation but that also outlined the role of the atmosphere in providing nutrition to plants. At the same time he also explained the role of minerals in plant nutrition.[20] Davy's main question, however, still remained. Agricultural chemists in the 1840s, including Liebig, were trying to understand how elements became incorporated into, and circulated through, plants and soils. They focused on nitrogen, carbon, and the role of humus in these interactions. Liebig was especially obsessed with the source of nitrogen for plants, calling it "the most important object of agriculture."[21]

Though scientific research in agriculture dates back to at least 1806 in Germany, Liebig's name in the 1840s became synonymous with "agricultural chemistry," and with the progress and usefulness that science could offer to agriculture.[22] Historian of science Margaret W. Rossiter remarks about Liebig's 1840 *Organic Chemistry* that "Liebig thus managed in less than two hundred pages to put together a wholly new synthesis of agricultural chemistry. It was probably one of the most important scientific books ever published and marks the beginning of a scientific revolution."[23] Liebig's new chemistry-based agriculture relied on inputs to the soil to replace the elements that were taken by crops. Well into the eighteenth century, agriculture functioned as more of a closed system, with nutrients being replaced with farm by-products such as manure, grass shafts, and fallow periods for fields. With the inception of growing food expressly for the market, most of the nutrients that would have reentered the farm system were taken away to distant markets, and less land was put to fallow. As described by the Marxist agricultural economist Karl Kautsky in 1899, replacing this deficit was the purpose of chemistry: "Chemistry is not only able to specify what these [lost] substances are, but can also synthesise the ones which the soil lacks, and which the farmer could not produce in sufficient quantity or only at excessive cost. Stable manure is not sufficient to maintain the equilibrium of modern agriculture producing for the market—especially a market which does not return the vast bulk of the nutritional material which it takes from the land."[24]

Liebig set up his initial laboratory in Giessen, Germany, under the patronage of Alexander von Humboldt, and in 1851 he founded a more enduring station in Mockern, just outside of Leipzig. This was the world's first official "agricultural experiment station," consisting of a 120-acre experimental farm and a small laboratory, emphasizing the need of field experimentation as well as chemical investigations that required a separate laboratory away from the field. While German and American scientists saw Liebig's experiment station as a successful "middle way" between science and practice, convincing most American farmers of the need for laboratory experiments in soil chemistry—in other words, dictating how they should farm—would be a challenge.

AGRICULTURAL SCIENCE COMES TO THE UNITED STATES AND SOUTHERN MICHIGAN

The first governmental agricultural experiment station in the United States was opened by the state of Connecticut in 1875, while Kellogg was laying ground for the new sanitarium building in Battle Creek. By that time, however, Michigan was already well established as a place with strong ties between government-funded scientific research in agriculture and educational outreach. Upon admission to the Union in 1837, the state government made a "geological" (though really more of an agricultural-suitability) survey of its lands. The benefits of such a survey were claimed to be the following: (1) It would disseminate knowledge of soils, (2) it would show how to correct deficiencies in soils, (3) it would reveal mineral manures, (4) it would accumulate information about destructive insects, and (5) it would disseminate knowledge of plants, both useful and noxious.[25]

The impulse by American government scientists to rationalize agriculture, as was already happening in Europe, occurred earlier and more strongly in Michigan than it did in most other states. Creating a catalog of problems and solutions to them—like deficiencies in soils, as well as entomological and botanical pests—was the beginning of an approach to farming that would increase homogenization of farming practices, styles, and biotic ecologies. Michigan's state agricultural college (now Michigan State University in East Lansing), which opened in 1855, was the first such institution founded in the country.[26]

After Connecticut, subsequent experiment stations in eastern and midwestern states opened throughout the 1880s and 1890s, with Michigan's opening in 1888. The role of the experiment stations was to "perform the experiments in cultivation which the individual farmer, lacking time and opportunity, could not."[27] Results from all kinds of experiments having to do with soil quality, necessary soil elements, and how to manipulate the balance of those elements through the use of farm inputs were then—ideally, at least—relayed to the station's constituent farmers. According to historian Alan Marcus, scientific agriculture was based most decisively on the activity of experimentation. An experiment "was grounded upon precision or exactness, a formalness lacking in casual observations. . . . [I]t was a systematically pursued observation undertaken according to and governed by an established series of ordered procedures and, as a consequence, reproducible."[28]

The experiment station brought agriculture into league with the rest of society's advances in manufacturing, transportation, and even health. As Marcus continues, the "professional scientific competence and the restoration of American agriculture, . . . experimentation marked the true scientific professional. . . . To experiment scientifically on the farm required the same means as those employed in the other sciences and in late nineteenth century America that meant access to a laboratory."[29]

The lure of the laboratory meant that between 1889 and 1900 the number of chemists employed nationwide at the agricultural experiment stations rose from 106 to 143.[30] To put this in perspective, in 1900 the experiment stations nationwide employed just fifty-five botanists, fifty entomologists, sixteen meteorologists, seven irrigation engineers, six biologists, and six zoologists. Accepted claims to proper practice in scientific agriculture, then, rested safely within the discipline of chemistry.

Liebig's program of agricultural science—while not without its critics—was eventually formalized in the United States by the aid of two congressional acts. The U.S. release in 1840 of Liebig's *Organic Chemistry* caused a great stir among farmers who were wary of theoretical knowledge, and who doubted Liebig's claim to guaranteed soil fertility. Yet he forcefully advocated for a "rational system of agriculture" based on "scientific principles" to increase soil fertility, the knowledge for which he claimed "we must seek from chemistry."[31] After decades of effort by a number of American agricultural scientists who had trained in Leipzig, the U.S. Congress passed two pieces of legislation that reflected Liebig's advocacy.

First, the Morrill Land Grant Act of 1862 provided funding for the development of agricultural colleges, the mission of which was to advance knowledge about crop growing and to train young farmers in progressive methods. And second, the Hatch Act passed in 1887 funded states to open and operate agricultural experiment stations.[32] The act's passage supported scientization in American culture, creating "a network of federally supported state stations devoted to agricultural investigation and experimentation."[33] When the act passed, systematic experimental work was already under way in Michigan, but no station was associated with the land-grant college.[34] The federal government's Department of Agriculture, which had just been created in 1862, promised that these stations would serve as the middle ground between agricultural practice and theory, and therefore consisted of both land plots and chemistry laboratories.[35]

These two acts signaled the political success of a growing interest among scientists and farmers in the role of scientific agriculture and its promise to systematize crop production, turning farm fields and the soil they contained into the products of laboratory experimentation. Chemists working at the state experiment stations asked how plants received nourishing elements (nitrogen, carbon, phosphorous, ammonia) from the atmosphere and the soil, and in what proportion these elements were required in order to maximize yields. The activities of people associated with these institutions represent an ideal—that is, how they thought the direction of agricultural practice should happen. While on-the-ground practice varied with each farmer's preferences, and while those preferences were sometimes at odds with the ideals, knowing the theoretical perspectives of agricultural "experts" helps us to understand an emerging type of agriculture that allowed for the mass production of grains.

The passage of these acts was largely the result of lobbying by the American agricultural scientist Samuel Johnson, who was convinced that the United States needed to replicate the research methodologies and funding structures from Germany, where he had studied.[36] Johnson had gone to Leipzig in 1853, and it was there that he became convinced that Liebig's was the best "middle way" between agricultural practice and theory. In the early 1870s Johnson and his student Wilbur Atwater led the drive that culminated in the Connecticut experiment station. Johnson also worked to show distrusting farmers the benefits of scientific agriculture (e.g., chemical soil and fertilizer analysis). His two texts—*How Crops Grow* (1868) and *How Crops Feed* (1870)—were used extensively for forty years after their publication in agricultural colleges and at the experiment stations. Johnson practiced and contributed to the development of soil physics, which included variables such as soil absorption, capillarity, permeability, cohesiveness, expansion, and retention of moisture. He believed that science and art, or theory and practice, were not to be separated. Science, however, always preceded art: "the farmer without his reasons, his theory, his science, can have no plan."[37]

WILBUR ATWATER: THE APPLICATION OF AGRICULTURAL SCIENCE TO HUMAN BODIES

Wilbur Atwater used the Department of Agriculture's experiment stations as a platform for instigating research in human nutrition, "a new field in which he faced no competition."[38] As someone who rose to power as the eventual director of the stations under the tutelage of Samuel Johnson, Atwater was armed with the ideas, tools, funding, and laboratories to extend the USDA's mission of improving agriculture to human bodies. In 1871, having just returned from two years of study in Leipzig and Berlin, Atwater was a professor at Eastern Tennessee University. In Germany he had studied with Carl von Voit, who like Liebig was a chemist interested in applying lessons from plant physiology and nutrition to humans. During this time Atwater traveled back and forth between Tennessee and Germany, "studying minimum daily protein and caloric requirements and the effects of such characteristics as age, sex, and occupation on nutritional needs."[39] Atwater then accepted a job at Wesleyan University in Connecticut, where he helped his mentor Samuel Johnson lobby for the creation of the agricultural experiment stations, using the argument that they were needed to combat widespread fertilizer fraud. Atwater and Johnson used this platform to argue that research unrelated to soil fertilizers—such as nutrition—was important as well. Atwater claimed of station scientists that "their chief work is the study of the broad and intricate questions of animal and vegetable nutrition, [and] the learning of the laws of animal and vegetable growth."[40]

Justus von Liebig saw that the twin processes of absorption and circulation that he studied in soils and plants were applicable to animal bodies as well.[41] He was concerned with the "respiratory metabolism" of animals, and showed that "the heat produced by the body could be entirely accounted for by oxidative processes, as opposed to . . . vital activity, such as electrical forces or nervous activity."[42] Underpinning these experiments was a concern with how much protein, carbohydrates, and fats were required for bodies to produce this heat, experiments guided by questioning the relationship between food consumption and bodily activity.

In the 1880s Wilbur Atwater picked up on Liebig's direction and began to apply the ideologies of rationalization and deliberation to food *after* it was harvested and eaten, culminating in the publication of a number of nutritional guides in the 1890s.[43] These guides were funded with the intention of maximizing the health of the working class and increasing the productivity of the nation.[44] If nutrition could become cheap and efficient, the working class would not have to spend excess time and money on food, a problem that was believed to threaten the functioning of the entire urban industrial economy. Importantly, these workers were less and less connected to the landscapes from which the food products came; agrarianism was well on the decline as the people who packed and transported food in cities was on the rise. For these workers, Atwater provided suggestions about which foods to eat, which were unnecessary, and which were affordable for the modern working body.[45] The roots of eating right—what Foucault might have called an alimentary form of "biopower"—in America began with Atwater, and is still strong today, as evidenced by the USDA's Food Guide Pyramid.[46] Displaced farm workers would become urban consumers in the new market-based food system, and their diets would suffer in ways that greatly concerned John Kellogg.

KELLOGG TRANSLATES ATWATER

Kellogg's style of quantifying the elements of food in his diet kitchens was taken from the work of Atwater. Nearly identical charts with categories of carbohydrates, protein, fat, and calories populate each of their publications, and Kellogg refers to Atwater as the source for this method. Kellogg demonstrated his commitment to food as a quantifiable medicine when he wrote that

> When the composition of a food is known, its calorific value, that is, the number of food units which it contains, may be easily determined. . . . By the use of these factors, it is easy to make a list of foods of known energy value per ounce, provided one has at hand a table showing the percentage composition of foodstuffs. Extensive tables of this sort are published by the United States Government. Bulletin No. 28 (Revised) of the Agricultural Department is particularly valuable. The data furnished by this bulletin have served as the foundation for the tables which are in use at the Battle Creek Sanitarium."[47]

The foundational document to which Kellogg refers—Bulletin No. 28—was first published by Atwater in 1896 under the USDA Office of Experiment Stations, and is titled *The Chemical Composition of American Food Materials.* Atwater brought the study of nutritive values of human food into practice in the United States via the Agricultural Experiment Stations at which he was employed. Following Liebig and Voit, he did so by quantifying the amount of refuse, water, protein, fat, carbohydrates, and ash for an incredible number of what he calls "American food products."

Atwater's dissection of a multitude of foodstuffs provided Kellogg with the format and the methodology to concoct the "perfect" diet for his patients. In Kellogg's adaptation of Atwater's presentation he shows the number of proteins, fats, and carbohydrates that a typical body needed at the sanitarium to stay healthy (table 1). In this table Kellogg breaks down the food products to be used for diet prescriptions, whereas Atwater used these three categories plus "refuse, water, and ash." Atwater defined *refuse* as "the bones of meat and fish, shells of shellfish, skin of potatoes, bran of wheat, etc." and *ash* as mineral matter that includes "potassium, sodium, calcium, and magnesium chlorides, sulphates, and superphosphates."[48]

Q: WHY MACHINES? A: APPROPRIATIONISM

Tractors, plows, and harvesters—many gas powered, and all of increasing size—dominated agriculture in this period, and they all found a market

Table 1 Height, Weight, Skin Surface, and Number of Food Units or Calories Required Daily

Body Size			Calories			
HEIGHT IN INCHES	WEIGHT IN POUNDS	SURFACE AREA IN SQ. FT.	PROTEIDS	FATS	CARBOHYDRATES	TOTAL
Men						
62	110.0	15.1	165	495	890	1650
63	115.5	15.6	173	519	1038	1730
64	121.0	16.2	181	543	1086	1810
65	126.5	16.6	190	570	1140	1900
66	132.0	17.0	198	594	1188	1980
67	137.5	17.4	206	618	1236	2060
68	143.0	17.8	215	645	1290	2150
69	148.5	18.2	222	666	1332	2220
70	154.0	18.6	231	693	1386	2310
71	159.5	18.9	239	717	1434	2390
72	165.0	19.3	247	741	1482	2470
73	170.5	19.7	255	765	1530	2550
74	176.0	20.2	264	792	1584	2640
Women						
57	78.4	11.9	118	344	688	1180
58	83.6	12.5	125	375	750	1250
59	88.8	12.9	132	396	792	1320
60	94.1	13.4	141	423	846	1410
61	99.2	13.9	149	447	894	1490
62	104.5	14.4	156	468	936	1560
63	109.3	15.0	163	489	978	1630
64	115.0	15.6	172	516	1032	1720
65	120.2	16.0	180	540	1080	1800
66	125.4	16.5	187	561	1122	1870
67	130.7	16.9	195	585	1170	1950
68	137.0	17.4	205	615	1230	2050
69	143.0	17.8	215	645	1290	2150
70	149.0	18.2	223	669	1338	2220
71	155.0	18.6	232	696	1392	2320
72	161.0	19.0	241	723	1446	2410

SOURCE: Adapted from John H. Kellogg, *The Battle Creek Sanitarium System: History, Organization, Methods* (Battle Creek, Mich.: Gage Printing Co., 1908), 123.

with farmers who were trying to sell grains that were in high supply. A Michigan farmer in 1892 describes the economic conditions of agriculture with regard to the use of machinery: "The use of improved machinery on the farm has increased the power of doing work of the producer manifold. . . . Prices for country products have, generally speaking, lowered during the last decade, which goes to show that production is still a good ways in advance of consumption."[49]

Low market prices through the 1890s and the ensuing social unrest were termed by late nineteenth-century political economist Edward Bemis as "the discontent of the farmer."[50] Farmers in Michigan profited not by selling a few high-quality, expensive products, but by producing as much as possible to sell many low-priced bushels of grain. For a strategy with such a low margin of error, where one failed crop could put a farmer out of business, technology updates that promised full, healthy crops that could be quickly prepared for market were very attractive.

The direct confrontation with nature sets agriculture apart from all other sectors of the modern economy. As noted by the twenty-first-century geographer Richard Walker, abstractions of the market do not easily match up with natural processes. "Making biological organisms jump through productive hoops has not always been easy, . . . and the growers have often had to turn to modern science in order to unlock nature's secrets—a process that continues in today's biotechnology revolution."[51] Agriculture has long confounded the free market economic system. Because of the necessary time and risk investments made by farmers to account for the nature of their product, food production is subject to dramatic boom-and-bust cycles that normally operate on a slower time scale than fluctuations in market demand. Farmers are unable to quickly respond to market demands because they must manage challenging variables such as rainfall, pests, temperature, and soil condition. Additionally, land itself is not part of a farmer's capital that can be bought and sold at a degree of flexibility that matches with fluctuations in the market. It is the "nature" in farming that has kept it distinct with respect to other factory-model production processes that emerged with the industrial revolution in the early nineteenth century, and is why subsidies have been a mainstay in the business. This is hardly a new observation. Edwin Willits, president of Michigan's State Agricultural College in 1885, captures this distinction of how farming is different than the other "mechanic arts":

> There is a radical difference between agriculture and the mechanic arts. A skilled artisan can select his material and apply his principle and construct a machine that shall be practically complete, perfect, and in accord with his plans and expectations. But no farmer can with certainty predict the outcome of his labors. What with the weather, untimely frost, difference in soils, blight, and mildew, uncertain germination, injurious insects, birds, and animals, coming unheralded and in unexpected character and numbers, noxious diseases to crop and beast, and the vicissitudes of seed-time and harvest, there would seem to be too much uncertainty in agriculture to call it a science, and too many elements of doubt to be solved anywhere by any one, much less in a school room by theorists.[52]

Agricultural science promised to eradicate, or flatten, these variables by systematizing each stage in the food-growing process, from seed selection to the depth of seed drilling to soil inputs to harvesting machinery. Willits doubted this promise of a universal solution, that all farms would benefit equally by off-site research. Instead, he represents a side of the debate that recognized the unique topographical and specific characteristics of each farm, and therefore the idiosyncratic, *non*reproducible knowledge of each farmer.

The productive limits of the land, and the incessant pushing of those limits by the industrialization of agriculture, have been described as "appropriationism." This refers to the process of transferring each component of agricultural production (e.g., seed improvements, fertilizing, or threshing) to specific sectors of industrial activity that take place *off the farm*, thereby removing a set of activities from farms that had previously contributed to a relatively closed economic and ecological system. The rise of guano importers, or agricultural machinery factories, for example, took nutrient circulation away from farms and put it in a global economic circulation. Critical observers of agriculture have claimed, therefore, that "this discontinuous but persistent undermining of discrete elements of the agricultural production process, their transformation into industrial activities, and their reincorporation into agriculture as inputs" are what they "designate as appropriationism."[53] This type of economic activity began in earnest in the mid-nineteenth century and hit full stride by 1890. For agriculture to be industrialized and food to be commoditized, the entire process of growing food had to be compartmentalized and broken down into all of its constituent parts. These processes were replaced—broadly—by machinery, chemistry, and biology, categories that roughly align with technological developments in agriculture through the nineteenth and twentieth centuries. Plowing, then, for example, became a separate, marketable process with the invention of the moldboard plow in the 1850s. It was an object that fit, unlike edible plants including wheat and corn, into a factory system that did not rely on seasonality, temporary labor, or natural disasters.

Due largely to generous federal land policies such as the 1862 Homestead Act, much of the upper Midwest was under agricultural land cover by the beginning of the Civil War. Integral to this phenomenon is the role that agricultural machinery played in "the opening of the prairies."[54] Prior to the 1850s the use of mechanical implements on midwestern farms was generally restricted because of the costs associated with purchasing and maintaining the machines. It was cheaper, if slower, to hire laborers. This changed when farm machinery manufacturing became more efficient, and the price of labor rose due to demands from other industries and the construction of railroads, both of which took away farmhands to higher-paying jobs. Between 1860 and 1900 the amount of money that farmers in this region invested in horse-drawn machinery—wagons, seed drills, broadcast sowers, mowers, cultivators, and threshers—nearly doubled, as it was a time when farmers learned to "farm sitting down."[55]

FROM EXTENSIFICATION TO INTENSIFICATION

Many factors contributed to the widespread adoption of machinery by farmers, though the most important force was economic. That is, for a midwestern farmer to stay in business in the second half of the nineteenth century, his yields per acre had to progressively increase. Many historians of agriculture attribute this necessity to the flooding of the grain markets that stemmed from the liberal land policies, exemplified by the Homestead Act. With more and more acres of prairie under cultivation, there was a surplus that not even global markets of the time could sustain. The solution that most farmers enacted to survive in a climate of ever-decreasing crop prices was to produce ever more crops. This required a shift in farming practices from extensification to intensification, a process that took place over the course of the century. At the beginning of the nineteenth century American farmers were more readily able to extensify their land holdings if they needed to produce more crop for subsistence or for market. But as population pressures grew and the nation moved toward the "closing of the frontier," farmers could no longer move onward or outward.[56] They had to increase the productivity of what land they already had. This, generally, marks the shift from extensive agriculture to intensive agriculture. Intensive agriculture, then, would become the dominant paradigm for American crop production that is still clearly visible today. Intensification consisted of soil inputs such as new forms of fertilizers, allowing for greater yields per acre, as well as new machines that allowed the same number of farmers to work larger tracts of land. In the language of progress, of course, this was all excellent in spite of the fact that many farmers were unable to keep up, and were forced to migrate to impoverished urban areas. As the contemporary political economist and cultural critic Thorstein Veblen put it in 1893, "Agriculture is fast assuming the character of an 'industry,' in the modern sense, and the development of the next few decades may not improbably show us, in farming as in other occupations, a continual improvement in methods and a steady decline in cost of production, even in the face of a considerably increased demand."[57]

The changing methods of growing, and declines in costs of production, were the cornerstones to the new, intensive form of agriculture, the efficiency of which is described by political scientist C.F. Emerick, who in 1896 wrote that "better methods of husbandry, the use of superior implements, specialization of agricultural production and vastly improved transportation facilities, whereby large areas of new lands have been brought under cultivation, have been indispensable to this increase in productive efficiency."[58]

Following Liebig, Johnson and Atwater led the charge of chemistry's role in agricultural intensification, which sought to separate, categorize, and rationalize every component of the process of growing food to maximize yields. For this scheme to work, farmers had to use machinery, and by the 1860s a widespread farm-implement-manufacturing industry had emerged, benefiting from the demands of the grappling farmers. Farming was divided into a group of separate procedures—plowing, sowing, weeding, harvesting, reaping, and threshing—and the mechanization of each of these processes was under way by the 1860s. Many of these machines appeared concurrently because of the "bottleneck" effect of the division of farm labor. The bottleneck effect occurred when one part of the farming process that was aided by a machine outpaced another part of the process, so that, for example, there may be acres of a crop thanks to a mechanical seeder, but no means to harvest the crop in the same low-labor, mechanized manner. For this reason the mechanization of most processes in food production tended to emerge contemporaneously.[59] As the twentieth-century agriculture economist Willard Cochrane put it, "advances in one aspect of farm production required advances in other aspects; if one farm operation became easier or quicker, it was of little use to the farmer unless he could speed up other aspects of the production process."[60]

THE NICHOLS & SHEPARD THRESHER, FABRICATED IN BATTLE CREEK

Large-scale agricultural machinery was the medium through which power was applied to fields in order to transform raw material into commodities. These technologies were the mechanistic arms of the ideology of systematization. The design and fabrication of some of the most widely distributed farm machinery in the United States for tilling, planting, harvesting, and especially threshing was done in Battle Creek.

The use of mechanical threshers in Europe between 1862 and 1895 increased more than the use of machines related to any other farm activity, outpacing sowers, reapers, rakers, and ploughs.[61] To thresh a cereal grain plant is to separate out its kernels—the part valuable to farmers and eaters—from the stalk, or husk. In the United States, the thresher was crucial not only because it raised profits by reducing the number of required laborers on each farm, but because it could work faster than even a large group of workers armed with flails (individual hand-held threshing tools). Flailing was normally done in winter months, providing job security for rural wage workers, but as markets constricted approaching the turn of the twentieth century farmers began to send their fall harvests to market immediately. The thresher, and especially the steam-powered thresher, made this possible. At the same time that it was industrializing farming, it was the machine perhaps most responsible for the surge in rural-to-urban migration in this period.[62]

As each region and subregion in the Midwest had its own crop specialization, so too did the manufacture of different farm machines have their own geographic centers. The center of manufacturing for threshing machines sold throughout most of the United States was in Battle Creek, Michigan. Two companies dominated this industry: Nichols & Shepard and Advance Thresher. Here I focus on the Nichols & Shepard corporation.

Nichols & Shepard was a foundry established in 1848 that specialized in plows, mill machinery, and farm equipment.[63] Like most farm-implement manufacturers of the time, John Nichols was a blacksmith who forged individual hand tools at the request of farmers.[64] In 1852 Nichols, realizing that the act of grain threshing had not yet been "appropriated" from manual labor to a machine, made his first thresher in an attempt to fill the niche and to industrialize a component of agriculture that had not yet been. As he and his business partner David Shepard continued to improve the design of their thresher, the machine grew in popularity throughout the Midwest, so that in 1876, when the company officially incorporated, its annual sales aggregate was $1 million. It was the largest factory in the city at that time, employing "200 hands, and consum[ing] annually about a million feet of lumber, 1,000 tons of pig iron and 500 tons of wrought iron."[65] The company covered at least ninety-seven acres with workshops that made over 1,500 threshing machines per year, which were exported on flat car trains to farmlands throughout the United States.

The Vibrator Threshing Machine was the company's most lasting and popular product (see figure 21). This machine made possible and encouraged the systematization of agriculture, disregarding micro-variations in plants brought about by seeds, topography, and soil. It worked by feeding stalks of grass—wheat, barley, and so on—into the front end (at left), at which point they would move through five vibrating plates (diagonal, toothed, in figure 22), separating the grain from the shaft, and catching the grain in trays below. Local resident Edith Butler described a mechanical threshing event in Allegan County, near Battle Creek, during an autumn in the 1890s:

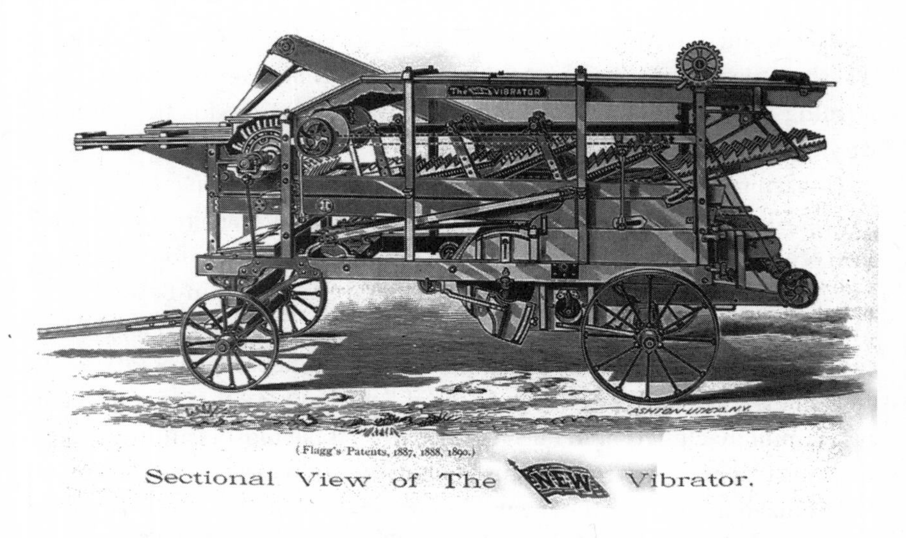

Figure 21. The Nichols and Shepard Vibrator Threshing Machine, ca. 1890. Image courtesy of the Bentley Historical Library, University of Michigan.

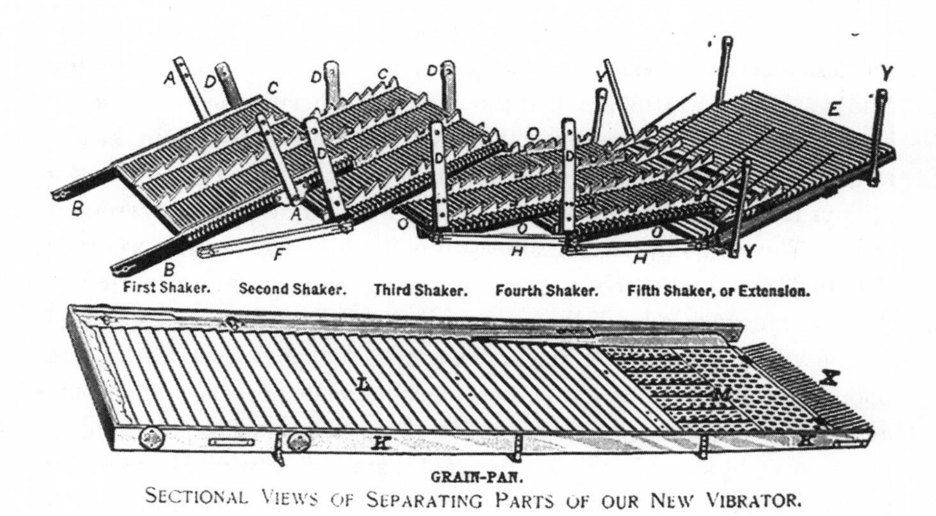

Figure 22. The Nichols and Shepard Vibrator Threshing Machine vibrating plates (above) and grain tray (below), ca. 1890. Image courtesy of the Bentley Historical Library, University of Michigan.

> The separator [thresher] was maneuvered into place at the barn door and the engine at the right distance down the lane so that a long, wide belt could run between a wheel on the engine and one on the separator. The water wagon was taken to the river and filled. A pile of wood was ready to stoke the engine and generate steam for power. The next morning, while the engineer built up steam in the engine, neighbors gathered, exchanging work for like help in their threshing. A man stood on the platform at the rear end of the separator to receive the bundles of wheat pitched to him. He push [*sic*] the bundles at the proper angle into the hungry machine. Out came the straw onto a carrier that conveyed it to a mow or stack. At the other side of the separator the grain poured into a sack. Two men alternated at grabbing the filled sacks and carrying them to the granary.[66]

Additional attachments allowed for the weighing and bagging of the grain, so that the farmer would not have to rake and collect grain from the bin. The ultimate effect of the threshing machine was "to produce the most marvelous separation of the grain or seeds, and most perfect motion of the straw through the machine ever known."[67] Variations in the quality of grains grown by farmers—whether variations in season or in the geography of individual fields—were not recognized by the Vibrator, homogenizing the task of threshing. Where special attention would have been paid to smaller, larger, looser, tighter, drier, or wetter stalks, the Vibrator enabled farmers to ignore, and perhaps not even notice these variations. The company claimed that a thresherman with a Nichols & Shepard Threshing Machine is "enabled to do thorough, clean threshing, no matter how wet or tough the straw, or what the condition of the grain or seeds."[68] This ability to ignore botanical variation is indicative of the changing nature of farm labor in the nineteenth century. The twin purposes of high yields and fast harvests were increasingly nullifying what had been, since at least the fifth century, a self-contained, self-reliant system, "a farming system of extraordinary strength and durability, conservative in the best sense of the term."[69] Those who opposed the advent of scientific agriculture feared exactly this loss of detail, local knowledge of landscape, which in the opposition's view was to the detriment of farmers' livelihoods and independence.

The threshers that were being mass-produced in Battle Creek were dispersed throughout upper Midwest fields. The labor energy that went into fabricating the threshers returned to Battle Creek in the form of train-car loads filled with wheat and other grains that were uniform in dimension and appearance, which would be transformed into health foods under the guiding hands of the Kelloggs. The manufacture of thresher machines in the very place where digestive systems were being reformed brings those machines into the geography of digestion. The rise of the health food industry, in which digestion was appropriated to the machines at the sanitarium, happened at the same time, and in the same town, in which farming became industrialized and appropriated.

OFF-FARM TECHNOLOGIES: MAIL-ORDER SEED COMPANIES

The "perfect work" done by the thresher meant that farmers responded to the machine's demands, curating increasingly homogenized fields, monocropped for the threshing style of the mechanics of the thresher. In order to use the Vibrator Threshing machine most efficiently, farmers needed fields of grasses to grow in the most uniform way possible. Making uniform fields began with finding uniform seeds that were guaranteed to take, and that were not susceptible to deadly pests and fungi.

When the USDA was founded in 1862, its two major directives were to disseminate information about agriculture and "to procure, propagate, and distribute among the people new and valuable seeds and plants."[70] On a continent with few native agricultural crops, the colonization of America depended largely on the importation of seeds from around the world, an endeavor that had persisted and was still in full swing at the time of the USDA's founding.[71] Even into the early 1880s, nearly one-third of the department's budget was devoted to global seed collecting and the distribution of those seeds to American farmers. When the Hatch Act was passed in 1887, officially sanctioning the agricultural experiment stations, "a regular program of exotic plant and seed distribution" was arranged through the station operators.[72] The government-funded project of collecting, categorizing, improving, and distributing seeds got mixed reviews from local farmers. As one anonymous Michigan farmer said, "clean fields and well fertilized land will render the necessity for changing seed much less frequent than some farmers now find it, if they wish good crops."[73]

Here he is speaking to the practice of changing seed varieties each planting cycle to ensure against insect and fungal pests. He is expressing a belief that reusing varieties from year to year is safe as long as fields are kept clean and the proper fertilizer is used. In a less progressive tone, however, he goes on to say that "quite a number of comparatively new varieties of wheat are being advertised this season, most of which have been tested to greater or less extent in portion of this State. . . . the question naturally arises, why are new varieties constantly coming up, and being sown by farmers, while old ones are gradually discarded until they are completely lost sight of?"[74]

When this question was posed in 1892 it was a very good one. Where were the seeds coming from? Beginning in the 1890s the source of those seed varieties for all sorts of grasses, fruits, and vegetables began to change from the government to private companies. At mid-century, because farmers were responsible for experimenting with seeds and there were no guarantees of their success, the government distributed seeds free of charge. Private seed companies at this time sold to the specialized markets of vegetable and flower gardens, and this despite a government distribution of approximately 2.5 million vegetable and flower seeds in 1861. Most of the substantial cash-crop varieties came from the farmers themselves, and the informal trading networks among them constituted the seed market. Then, in 1883, the American Seed Trade Association was formed as a response to the heightening importance of California's fruit and vegetable industry, an industry in which private seed companies could compete with the government. The Secretary of Agriculture in 1893—J. Sterling Morton—recognized the trend to privatization and tried to end the free distribution of seeds, claiming that the distribution of "new, rare, valuable, or other seed should be left entirely in the hands of the branch of industry to which it lawfully belongs."[75] While it took some time for the government to stop distributing free seeds, the penetration of private seed industry began in the 1890s.

Buying seeds that were the best suited for their particular soil, climatic, and disease conditions was part of the technological revolution in agriculture that is so often overshadowed by the story of farm machinery. Assigning all the different activities it takes to grow food to a certain set of machines was only one way to industrialize agriculture. The other way was to commoditize the seeds themselves, making constant improvements and variations for sale. The relationship between modern science and the rapid turnover of wheat (and other seed) varieties has its roots in the late nineteenth century.[76] When farmers started ordering seeds from warehouses that cataloged and archived seed varieties, they were reshaping their fields in the image of the factory-like precision of seed warehouses themselves. To make fields that were full, healthy, and uniform at harvest time, farmers increasingly relied on companies like the Harry Hammond Seedsman distributor in Fifield, Michigan (see figure 23). Hammond claimed that he had the fastest-maturing vegetable seeds, and the seeds that would produce the greatest yield, an attractive claim to farmers who were under economic pressure.

Connecting this distributor to the Battle Creek Sanitarium means that we first have to know what exactly was served at sanitarium meals. Figure 24 is an inventory of the foods consumed at the sanitarium over the course of a typical year. The source of these foods depends on the amount consumed, and the climatic growing conditions they require. Many of the fruits and vegetables, like apples, tomatoes, grapes, berries, beans, peas, and potatoes, were products of the sanitarium itself. As sanitarium historian Patsy Gerstner put it, "Much of the [sanitarium's] food was either grown in on-site greenhouses or purchased under contract from farms approved by Kellogg."[77] There was even a sanitarium mill, where grains were ground into flour.[78] The sanitarium also operated a number of farms off campus. By 1900, for example, the sanitarium had acquired nearly four hundred acres of farmland near Goguac Lake, providing the sanitarium with "a constant supply of milk products, eggs, fruits, and vegetables."[79] The use of the greenhouse and the farms near Goguac Lake was important to Kellogg; he was wary of foods that were shipped in from other places that did not meet his standards of cleanliness. Kellogg wrote that

> the use of lettuce, celery, and some other fresh vegetables as supplied *in market* is always attended, as Metchnikoff has clearly shown, with more or less risk of parasitic infection because of the careless use of night soil and other fertilizers by market gardeners. Such products, when not obtained from the Sanitarium gardens or greenhouses, are always sterilized in the kitchen by immersion in boiling hot water before serving. It is a help to the invalid's appetite to know that the table delicacies placed before him are thoroughly clean.[80]

Figure 23. The main floor of the seed-shipping department of Harry Hammond Seedsman Distributor. Reproduced from "Annual Catalog of Harry N. Hammond, Seedsman," 1900 (Fifield, Michigan). The catalog was published by the Review and Herald Publishing Co. in Battle Creek. Image courtesy of the Bentley Historical Library, University of Michigan.

Apples	1,959 bu.	Berries	37,632 quarts.
Apricots	1,418 lbs.	Berry juice	1,625 quarts.
Bananas	1,392 b'dles.	Beans	11,042 lbs.
Dates	2,346 lbs.	Peas	11,684 lbs.
Figs	7,246 lbs.	Potatoes	4,585 bu.
Lemons	477 boxes.	Eggs	44,643 doz.
Oranges	534 boxes.	Nut foods	1,374 cases.
Peaches	1,016 bu.	Malted Nuts	5,882 lbs.
Pears	212 bu.	Meltose	14,242 lbs.
Plums	279 cases.	Buns	5,542 doz.
Prunes	5,871 lbs.	Bread	67,698 loaves.
Tomatoes	16,464 quarts.	Granose biscuit	6,821 lbs.
Grapes	20,515 lbs.	Water breads	15,514 lbs.
Grape juice	8,750 quarts.	Flaked cereals	332 cases.
Apple juice	29,524 quarts.	Granola and granuto	2,688 lbs.
Pears (canned)	63 cases.	Zwieback	27,877 lbs.
Peaches (canned)	9,072 quarts.	Cereal coffee	3,137 lbs.
Plums (canned)	1,788 quarts.	Gluten meal and flour	1,566 lbs.

Figure 24. Inventory of annual consumption at the Battle Creek Sanitarium, by food, 1908. Image reproduced from John H. Kellogg, "The Battle Creek Sanitarium System: History, Organization, Methods" (Battle Creek, MI: Gage Printing Co., 1908), 138. Image courtesy of the Louise M. Darling Biomedical Library, History and Special Collections for the Sciences, UCLA.

Figure 25. Hammond Seedsman mail-order department, 1900. Image courtesy of the Bentley Historical Library, University of Michigan.

Figure 26. The Hammond Seedsman manufacturing plant, 1900. Labels on the drawers of the seed warehouse refer to this growing environment of greenhouses, not that of a farm field. Image courtesy of the Bentley Historical Library, University of Michigan.

On the right side of figure 25 are drawers in the Hammond mail-order department that reach about fifteen feet toward the ceiling. In these drawers are housed the seed varieties of plants ranging from potatoes to cabbage, asparagus to strawberries, and lettuce to oats, to name a few. The inset in the upper right-hand portion of the image shows women filling the orders by retrieving the seeds from their appropriate home and "packeting" them for shipment. It is unknown, but likely, that the sanitarium ordered seeds from this Fifield, Michigan, distributor, as the Hammond catalog was printed by the Review and Herald, the Adventist publishing house in Battle Creek. Hammond's company undertook the business of cataloging nature into different cells, each one with a label, yield measurements, maturing speed, and, of course, price. As sociologist Bruno Latour describes in his discussion on the "circulating reference," botanical varieties took on this new commercial meaning as objects in a market. The labels on the outside of those drawers in the seed warehouse became referents to particular organisms, experiments, and soil qualities from which the seeds came (see figure 26).[81]

Fruits and vegetables grown at the sanitarium farms and greenhouses—which would reform the shape of digestive systems—were connected with this seed industry at the Fifield seed farm. When John Kellogg talks about the constitution of bodily organs and their cells that are made through food consumption, then materially and symbolically the digestive systems of the sanitarium guests were made up of this privatized and cataloged germplasm. The digestive system was part of an assemblage that included a "bookkeepers'" view (Latour's term) of the natural world. But in addition to being a part of this material assemblage, digestive systems at the sanitarium were also privatized in much the same way that seeds were privatized because they fell under the expensive gaze of Dr. Kellogg. He, like the companies that made and sold seeds, was appropriating a part of the body, the digestive system, to be able to scrutinize it with an amount of detail that allowed him to concoct all sorts of treatments and cures. Opening the "black box" of digestion was profitable to the Kelloggs in the same way that opening the black box of seeds was profitable to private, "appropriating" seed companies.

6 Breakfast Cereal in the Twentieth Century

> That the eating habits of the American public have been materially modified is evidenced by the fact that thirty to forty carloads of toasted flaked cereals are being eaten daily under various names in the United States alone, and the consumption is steadily increasing and is rapidly extending to foreign countries.
>
> —John Kellogg, 1908[1]

The story of breakfast cereal changes dramatically after William Keith Kellogg—the younger brother of John Kellogg—gained control of the Sanitas Nut Food Company, which had been associated with the sanitarium from 1890 to 1908.[2] There has been much to tell and retell about the story of breakfast cereal after Will Kellogg transformed the boutique foods of the Battle Creek Sanitarium into a worldwide food corporation.

During the first decade of the twentieth century Battle Creek was home to an impressive number of factories, including steam pumps, breweries (including Anheuser-Busch), flour mills, furniture makers, printing presses, farm machinery fabrication plants, and health food manufacturers.[3] At this point cereal companies still represented only a part of the overall industrial economy in Battle Creek.[4] Food companies, while not yet the mainstay of the city's economy, rapidly multiplied with the growing success of the Battle Creek place name, which was becoming a code word for health food throughout the nation. Not surprisingly, between 1900 and 1915 over eighty cereal companies came into and went out of business in Battle Creek.[5] The American Pure Food Company, Battle Creek Breakfast Food Company, Battle Creek Cereal Food Company, Battle Creek Flaked Food

Company, Battle Creek Food Company, Battle Creek Food Products Company, and Battle Creek Pure Food Company are only a sampling of the seemingly endless variations of business start-ups that used the geographical indicator of Battle Creek in an attempt to sell their version of health foods to a national market for the new trend in breakfast eating.[6] They sold products with names as equally strange sounding as those used by the Kelloggs, such as Ce-Re-O-La, Egg-O-See, Flak-Ota Flaked Cooked Oat Food, Grain-O, Grape Sugar Flakes, Malta Vita, Navy Bean Cereal, and Zest Flaked Cereal. These companies mimicked the products produced not only by Kellogg, but also by Kellogg's chief contemporary rival, Charles Post. Kellogg and Post would together, through the early decades of the twentieth century, give Battle Creek its present-day title of "Cereal City." The "imposters," as they were called by the Kelloggs, were so numerous that in 1905 the famous cartoonist for the *Chicago Tribune,* John T. McCutcheon, published a satirical cartoon addressing the issue.

Surrounded by over one hundred competitors, how did Kellogg's company emerge as the most prominent? Starting in the mid-1890s John Kellogg could not keep the Sanitas Nut Food Company profitable. He was uninterested in developing a single product and then mass-marketing it, as the hundred others were trying to do. William Kellogg's biographer Horace Powell points out, for example, that "because of medical ethics, the Doctor refused to allow his name to be associated with the foods. . . . The Doctor was a conservative in business matters and was opposed to large-scale national advertising, permitting the new flakes and other goods to be advertised only in the *Good Health* magazine, in the Battle Creek *Idea,* and by direct mail to former patients. These restrictions chafed at the innate and developing business talents of Will Kellogg."[7]

While it may be true that Battle Creek "brought the advertising business out of its infancy," it was certainly not due to the work of John Kellogg.[8] By 1902 Sanitas was losing money, prompting William Kellogg to call his doctor brother "the best disorganizer in the world."[9] How the Kellogg brothers emerged from this is a story of branding and advertising that began when William Kellogg took over the business side of the sanitarium's food company.

Recognizing that hoards of competitors were capitalizing on the reputation John Kellogg had developed at the sanitarium, William understood that the health clinic and the food company should be moving in different directions. John Kellogg was under pressure from the American Medical Association (AMA) to avoid excessive commercialization, a warning that he took seriously since he was already under a watchful eye due to his "unorthodox" practices of hydrotherapy and massage.[10] But William consistently encouraged branding the Kellogg name, a claim to authenticity that would help them compete with the other food companies emerging in Battle Creek. William threatened to leave many times given his subservient role to his brother, but he felt committed to aid the rebuilding process after a 1902 fire burned the sanitarium to the ground.

John Kellogg contentiously posited that to save the reputation of his health establishment in the eyes of the AMA, it would be "best to dispose, if possible, of the toasted corn flakes, so as to relieve the Sanitarium."[11] After years of tedious courtroom litigation between the brothers, in 1906 William won control of the Sanitas Nut Food Company, separated its affiliation from the sanitarium, and renamed it the Battle Creek Toasted Corn Flake Company. The use of corn to make flaked cereals was less than ten years old at this point, and focusing the company's identity on it was risk considering that consumer preferences were centered on wheat. Under this new name, a commodity was born, and with it a new essence of flaked cereal. By 1908, under William's supervision, Kellogg's Corn Flakes had completely severed its ties with the health food industry created by John Kellogg. William Kellogg removed the image of the sanitarium from the box and replaced it with his signature, claiming simplicity and taste over the curative, digestive properties of flaked grains.[12]

WHY CORN?

William Kellogg's decision to use corn instead of wheat is partly a reflection of the changing nature of the agricultural industry. As early as 1860 the general structure of food production that we know today as the corn-based agricultural economy was taking shape in the United States. Changing tastes on the urbanizing East Coast created a large demand for meat, and it was in the Midwest that much of the pork was produced to meet this demand. In 1860 the rail networks had not yet been fully developed, so farmers would sell their grain to distributors who would then ship it on riverways to Cincinnati. It was here that some of the earliest prototypes of feedlots were being developed, earning Cincinnati the nickname "Porkopolis." Even at this early point in the development of the U.S. agricultural economy, corn had already taken a role as a fattening agent for meat-providing animals.[13]

In 1860, most of the corn in the United States was grown in the southern part of the Midwest and into the south. Tennessee, Kentucky, and southern Indiana led in corn production. It was not until the 1880s and 1890s that the great corn migration to the north began. While farmers in upper Midwest states such as Michigan, Ohio, Illinois, Iowa, and Wisconsin had been growing corn as well up to that point, the landscape was mostly dominated by wheat. In this period a relatively large number of farmers in the upper Midwest were still operating under the new husbandry model of agriculture, using part of their land for subsistence and part for market commodity production. By 1900, however, after a number of depression years in the 1890s that were marked by "the discontent of the farmer," farmers increasingly found themselves economically forced to plant their fields for market.[14] Farmers who did not maximize yields of the crop that had the highest market value found it nearly impossible to make enough money to buy the necessary goods and services to keep their farms in operation, let alone to buy things that families needed for everyday life.

Understanding why farmers in the upper Midwest began replacing wheat with corn means also understanding the region that neighbors the upper Midwest: the Great Plains (including the Dakotas, western Minnesota, Nebraska, Oklahoma, and Kansas). Wheat farmers realized that with cheaper land prices they could migrate to the Great Plains and have bigger land holdings. In the mid to late nineteenth century, it was economically advantageous to mass-produce wheat in these areas because of the availability and cost of land there. This economic factor that pulled wheat westward, however, left a hole in the upper Midwest. The market value of corn dictated to land managers, owners, and farmers that they should plant corn to turn a profit in these regions, since if they kept planting wheat they would be undercut by the Great Plains farmers. And plant corn they did. It was the categorization of corn as something with potential exchange value at the Chicago Board of Trade—ultimately to feed the beef and pork industries—that flooded the upper Midwest with corn and turned it into the "ocean" of undifferentiated commodity that we still see today.[15]

In anthropologist Sidney Mintz's geographical history of sugar, he tells how it played a significant role in feeding the new industrial working class, in line with Atwater's research. But Mintz also tells how the relationship between metropoles and their colonies changed with the way sugar was tasted by the powerful.[16] Transformed from a rare and un-sought-after medicine to a luxury item, and finally to ubiquity, its path is in many ways like that of breakfast cereal as it emerged from Battle Creek. The methodological lesson of Mintz's story is that we can use the consumption of foods as markers for changes in the ecology of that food's production landscape. Due to the changing meaning of sugar at the consumption end, a motivation for overseas agricultural experiments emerged at the production end, dramatically affecting landscapes and ecologies throughout European colonies, especially in the Caribbean. At the consumption end, the item of sugar had been transformed from a luxury to a necessity, which "embodies both the promise and fulfillment of capitalism itself."[17] Kellogg's breakfast cereal story can be thought of in similar terms. Beginning as a medicine with low-volume production, the product was latched by William Kellogg to new practices in mass-marketing strategies, and, of equal importance, to a new form of capitalist agricultural economy that drove the price of corn down and its availability up, responding to and encouraging changes in consumer tastes. It is no small coincidence that the single most important factor in the rise of breakfast cereal as a popular food commodity in the twentieth century was its marriage with sugar. For moral and medical reasons, John Kellogg remained vehemently opposed to augmenting flaked grains with sugar. It was one of the major points of contention he maintained with his brother, Will, as the industrialization of corn flakes intensified. Breakfast cereals in this lineage today are little more than carriers for sugar, serving as evidence for the winner of the Kellogg brothers' dispute.

ADVERTISING

Under William Kellogg's new control of the food company, and with a steady, cheap supply of corn, advertising would be another major mainstay of the business' success, with unprecedented percentages of revenue devoted to marketing. William's advertising expenditures first exceeded $1 million in 1915, a figure that would only increase. His innovations in advertising were based on ruthless aggressiveness, risk taking, and enormous capital investment. Causing the biggest stir was the 1910s ad that asked housewives to "wink" at the grocer, presuming he would know she wanted Corn Flakes, using sexual desire to more fully undercut John Kellogg's puritan sensibilities. Another long-standing campaign that daringly used sex appeal was a series of ads that pictured the "Sweetheart of the Corn."

Other revolutionary campaigns included sending young women dressed as the Sweetheart of the Corn to food shows where they would pass out samples; mail-in baby picture contests; traveling canvassers who announced the arrival of Corn Flake shipments to towns; premiums, games on the box, cut-outs, and package inserts for kids; contests for grocers; men dressed up as ears of corn and cereal boxes wearing the Kellogg's colors of red, white, and green; and ads that used reverse psychology, begging people to *stop* eating Corn Flakes so that their friends and neighbors could have some.

Most extravagantly, perhaps, Will Kellogg erected a large electric sign in New York's Times Square with tears falling from the cheeks of a boy whose Corn Flakes bowl was empty. By 1931, as radio became nearly ubiquitous in American culture, extensive use of radio advertising was seized by the Toasted Corn Flake Company, using the voice of the "Singing Lady" on NBC's station.[18] Then, with the advent of television advertising, Kellogg's was again at the forefront of innovation. In a 1950s television advertisement for Corn Flakes, the early *Superman* actor George Reeves promoted the benefits of eating the breakfast cereal.[19]

THE PLACE NAME OF BATTLE CREEK

The "Sweetheart of the Corn" advertisement shows the long-standing attachment to place that has been used by the Kellogg Corporation in its self-representation. In the late nineteenth and early twentieth centuries, the place of Battle Creek was a significant part of the authentication and differentiation process for the Kellogg Company's food products. Just like

W. K. Kellogg's iconic signature on cereal boxes, Battle Creek—imagined as a place of health and healing—made the Kelloggs' foods recognizable and purportedly superior to imitations. This book has delved into the backstory of a food product rooted in place. The backstory presented in these chapters demonstrates how the consumption and assimilation of food into the body was dependent on the regional technologies surrounding the food's place of production.

One of the most amazing things about this story is that John Kellogg was able to galvanize and channel a variety of popular and scientific concerns—about nutrition, disease, city life, and agriculture—and transform those concerns into artifacts—health foods. William Kellogg was able to take those artifacts and transform them into commodities, commodities that represented a solution, a way to act responsibly for one's health and longevity, but also, and more importantly for William, a way to satisfy a desire for something safe, economical, and convenient. The United States Pure Food and Drug and Meat Inspection Acts were both passed in 1906, months after Upton Sinclair's muckraking exposition of the American meat industry—*The Jungle*—was published.[20] Thanks in large part to Sinclair, public concern over food adulteration and sanitary conditions at the source of food processing was widespread and powerful. William Kellogg's Waxtite packaging, clean, tidy boxes, and the tradition of health attached to the Battle Creek place name assuaged those fears, and were responsible for much of the early success of the Toasted Corn Flake Company. It was the way John Kellogg conceived of the stomach that turned food into health food, and thus helped his brother William Kellogg market their grain-based commodity food cures nationwide. In essence they were selling a vision of the process of digestion inside the body that was invented *in situ* in Battle Creek. This was a body that could be ameliorated by consuming their mass-produced health food products, which would, in John Kellogg's view, relieve disease and prepare the body for enlightened living on earth and in heaven.

FOOD PROCESSING REVOLUTION

In addition to its contributions to the advertising industry, the Kellogg Company revolutionized mechanical food processing. The rollers in the sanitarium—where the first "predigested" flaked cereal was made (see figure 12)—were the beginning of what would become a progression of food-processing machinery that met the developing needs of the Kellogg corporation's changing product innovations. These innovations spread to other food manufacturers, foreshadowing a century that would see food markets change from purveyors of mostly perishable, local products to distributors of prepackaged, preserved, "sterile," boxed products that could be transported long distances.[21] As William Kellogg's chief biographer Horace Powell put it, "Look at any type of food production, particularly processing, and it will be seen that much of the impetus came from Battle Creek."[22]

Oftentimes William Kellogg would invent new machinery concurrently with the design of new health food products, implying that if a hopeful new product could not be mass-produced with a relatively simple set of machines, it would not be developed at all. The ability of a food design to fit into the range of "affordances" that technology could provide in large part determined whether or not it would come to exist.[23] William Kellogg was convinced that the single most important aspect of mass-producing foods is keeping them fresh. If foods were to be shipped and sold to customers covering a wide geographical expanse, then maintaining the crispness and flavor of baked and malted grains would require technological developments in packaging. The most successful development came in the form of the "Waxtite" package, which was eventually contracted out to the U.S. government for their "K-rations" during World War II.[24]

PAIRING MILK AND CEREAL

Like sugar, the now-common union of breakfast cereal with milk is not a coincidence. While cereals were initially made to physically clean out, and morally purify, the body, milk was concurrently being made as the single "perfect food" for Americans.[25] Claims about the perfection of milk began as early as 1834, with English physician William Prout's *Chemistry, Meteorology, and the Function of Digestion.*[26] Believing that milk could be the single savior to problems of nutrition was, however, not obvious. As physicians and public health workers discovered that milk could carry germs for diseases such as typhoid fever and diphtheria, sanitation became a paramount concern surrounding the production and consumption of milk. By the 1890s, some thirty years after its invention, pasteurization had become a key process in the certification of milk for a product that was riddled with a paradox that came to be known as "the milk question," which was "How do we provide ourselves with the necessary food without being poisoned by it?"[27] If milk was to retain its status of perfection, it would have to be squeezed into the prevailing ideas of a pure, clean, hygienic food.

Historical roots of the confusion inherent in "the milk question" paralleled John Kellogg's inconsistent discourse on milk at the sanitarium. He wrote, for example, that cows with "impure" blood produce milk that carries the impurities of "foul air" from the stable: "One half the deaths among the young are directly traceable to poisonous milk; and yet thousands of people, especially in our large cities, are daily exposing themselves and their children to the possibility of fatal poisoning."[28] Yet five years later he claimed that "milk contains the different elements of nutrition in proper proportion, and will sustain life for an indefinite period. . . . It is, indeed, with rare exceptions, a most wholesome food for persons of all ages."[29] Transitioning milk from a potential poison to "the perfect food" was done with such precision by urban reformers, medical professionals, and dairy cooperatives that the change was reflected in Kellogg's rhetoric, and in the contents of sanitarium guests' stomachs.

The creation of perfect milk went hand in hand with Kellogg's philosophy of health. Through one of the few advertising- and government-led public health campaigns of its scope that predated the humongous private advertising campaigns of Kellogg's cereals at the turn of the century, milk had been ingrained as a healthy food that everyone should consume to stay healthy. Given Kellogg's repulsion toward meat and animal "filth," it comes as a surprise that even his often-contrarian dietetics frequently included the ever-popular milk. Even other dairy products, such as cheese, did not win the favor of John Kellogg. His adaptation of the "perfect food" story that has been so ubiquitous in the American dietary rhetoric regarding milk not only shows the enshrinement of milk's status in the stomachs of Americans, but also helps us to understand the pairing of milk with whole grain flakes of wheat and corn. In sum, Kellogg thought that if you could have one perfect food—milk—why not pair it with his own perfect food—flaked cereal?

ANTIMODERNISM AND MODERNISM

John Kellogg's mission to concoct and serve these perfect health foods to large constituencies in Battle Creek was not met without cultural opposition. By the end of the nineteenth century there had emerged a strong antimodernist sentiment in American culture. Historian Jackson Lears describes this sentiment as a "recoil from an 'overcivilized' modern existence to more intense forms of physical or spiritual experience supposedly embodied in Medieval or Oriental cultures."[30] *Antimodernism* is a term that encapsulates a reaction against tendencies of rationalization, urbanization, efficiency, and secularization; in short, it worked against what we see in Kellogg's understanding of digestion, Battle Creek's modernization of its sanitary sewer system, and the scientization of agriculture at the experiment station in Lansing. Pure, ordered, efficient, time- and purpose-driven places were viewed with mistrust by antimodernists. Though antimodernism is perhaps better conceived as one of many types of modernization, it remains that a critical interpretation of Kellogg's activities did exist at the sanitarium; think again to the example of the "meat speakeasies" from chapter 2.

Nonetheless, the trends of both modernization and antimodernism are represented simultaneously at the sanitarium in the years preceding the turn of the twentieth century. John Kellogg's quantification of diet and digestion are primary examples I have used to highlight how modernization was manifested at the sanitarium. In addition to measuring the composition and ratios of food elements such as fats, carbohydrates, and proteins, John Kellogg developed time charts for exactly how long food should remain in the body (figure 16). However, these modernizing trends occurred at the sanitarium in the context of a paradigm shift that was moving away from more traditional, environmental explanations of health. The 1890s, perhaps more than any other decade, is a period in which discourses from these two large trends intermingled with relative ease. Thus, sentiments of nostalgia for nonrationalized foods would not have been out of place. This is especially believable for the sanitarium's elite urban clientele, who would have been exposed to enough "modernism" to support a backlash in the first place. A trip to the western lands of semi-rural Michigan was a movement back in time, for many guests, to a perceived simpler and better time. But this movement back in time remembered only the best attributes and discarded the undesirable parts, as nostalgia often does.[31] It made sense, then, that Kellogg's scientific, modernizing practices were able to meld so effectively with the escape *from* modernity that the sanitarium also offered. Wood splitting, jogging, camping, and boating, for example, were all activities undertaken between sessions of gastric juice analysis, colonic machine enemas, electric and water therapy, and prescription meals.

ANTIMODERNISM AND THE NONNATURALS

Another helpful way to understand antimodernism at the sanitarium is to look again at Kellogg's medical philosophy, which partly drew from ancient beliefs. The doctrine of the nonnaturals reflects ideas about health, and was an inspiration for Liebig's work—the study of metabolism. This doctrine was best explained by James Whorton, whom I quote at length here:

> Non-naturals was the somewhat confusing term that had been used for centuries to categorize the various components of behavior and environment that one had to regulate so as to live in accord with nature. In the classic formulation, they comprised air, food and drink, sleep and watch, motion and rest, evacuation and repletion, and passions of the mind. Derived from everyday experience and common sense, the non-naturals simply recognized that people needed a wholesome atmosphere to breathe, adequate food, sleep, and exercise, regular elimination of waste, and emotional equanimity. That was the meaning of the word hygiene in the nineteenth century: adherence to the doctrine of the non-naturals.[32]

The six categories of the nonnaturals represent the flow of matter through the body, and therefore relate the consciousness of a material connection between humans and the environment. Liebig was among the first to apply a systematized, quantified, mechanical, causal language to this material interchange, which would influence not only the agricultural experiment stations and human nutrition, but also the emergence of the field of ecology.[33] As classically defined in the doctrine of the nonnaturals, people were unavoidably exposed to each of the six characteristics at all times. Since they were unavoidable and recognized as activities that everyone did, it became the role of the physician to restore any imbalances in any of the six main categories. This is how medicine was practiced, and how health—and unhealthiness—were defined. Imbalances of the nonnaturals were used as an explanation for someone who felt ill or in pain, but were also used as moral guidance for people who were afraid of becoming ill or in pain. Dietetic, sexual, and exercise directives, for example, fall into the latter use. Preventative and practiced medicine came together in the use of the nonnaturals. In the middle of the nineteenth century the language of the nonnaturals began to be replaced by the language of science. While the ideas remained, a rationalized hygienic philosophy, often marked by quantification and microscopic analysis, began to twist the ideas of the nonnaturals into medical specialties that would gain popular acceptance by the end of the century.

It was the habits of modern urban life that were wrecking the stomach. What makes Kellogg's story so interesting, though, were the moments of ambivalence. He wanted to cure at once with a philosophy of "nonnaturals" and with a philosophy of modern science. This ambivalence is reflected in the analysis of H. Thompson Straw, a 1930s urban geographer who wrote about Battle Creek: "The sanitarium's reputation, which has helped to make Battle Creek so well known, is not based primarily upon the practice of a specialized surgery and medicine, but rather upon what is termed 'healthful living.' This includes proper rest, food, and exercise. It is, therefore, the many minor and not the major ailments of mankind which the sanitarium specializes in treating, although it does maintain a well-equipped surgical and medical staff."[34]

Whorton further normalizes Kellogg's attitude: "one of the distinguishing marks of Western culture since the Enlightenment is precisely this *ambivalence* about the relationship of health to civilization."[35] Kellogg promoted what he called a hygienic philosophy of health, which rested on the notion that once disabused of the ignorance and errors of living, patients would find an (ambiguous) combination of success, happiness, and longevity. The genealogy of the hygienic philosophy can be drawn much further back in time. As one scholar pointed out, the notion "that ignorance and error are largely responsible for man's woes, including most of his physical ailments, is an ancient doctrine."[36] Placing the fault of illness directly on the individual empowered sanitarium patients inasmuch as it gave them control over their own destiny; it was not a matter of succumbing to the doctor's tools. If one could simply follow the dietary and therapeutic programs outlined at the sanitarium, they would direct their body and spirit to goodness.

MOVING BETWEEN LANDSCAPES

In examining what happens between consumption and production I look at the digestive system, the sanitarium building, the city of Battle Creek, and the agricultural hinterlands of southern Michigan. Moving among these places means considering what it is that moves between them. What parallels in the material world justify the conceptual movements that I propose? French philosopher Michel Serres describes the movement between places as being always parasitic. Noise, or static information, disrupts order in any given place and allows for the emergence of (temporary) order in another place.[37] Applying Serres's term here, what were the "noisy" phenomena in Kellogg's ideal digestion process? What were the inevitable moments of chaos that catapult us from one place of the story to another?

Asking what happens between consumption and production implies a spatial journey between the digestive system and the places affected by it. The story in this book is spatial rather than temporal. It is a mapping of the mechanical implements in each of their contexts. The mechanical implements chosen for examination are those that bring us from place to place to follow from consumption to production. The implements, or technologies, all in one way or another affect how digestion works at the sanitarium. Landscapes cannot be separated from the technologies that produce them.[38] It is in this sense that I use the term *landscape*—as an artifact of a particular set of tools, of a particular method of altering and creating environments. There is a congruence between the selected places in the geography of digestion. This congruence is found in the manner in which its environments are altered. Immediately proximate to the body is the environment of the sanitarium building. In the building are housed technological extensions that make digestion possible. Further removed spatially is the waste disposal system for the city of Battle Creek. Here we see that the entire underground urban infrastructure was created to make a sanitary city. The underground of the city was materially connected to digestion at the sanitarium, but the look of the new sanitary city is also connected to the new body produced by the new digestion. There are, therefore, both material and metaphorical links between sanitarium bodies and the surrounding landscapes. Returning to the idea of noise, or static, the question arises: at what exact points when we move between consumption and production do the ruptures occur? What is the break in the system that requires us to switch our gaze from the stomach to the sewer, for example?[39] The metabolism of the body must be extended out to the metabolism of the built (including agricultural) environment. The notion of metabolism and circulation of waste through bodies and environments is a nineteenth-century concept that described "material exchanges between organisms and the environment as well as the biophysical processes within living (and non-living, i.e. decaying) entities."[40] In sociologist Bruno Latour's formulation of actor network theory, he states that "we have to lay continuous connections leading from one local interaction to the other places, times, and agencies through which a local site is made to do something. . . . [W]hen you put some local site 'inside' a larger framework, you are forced to *jump*. . . . What would happen if we forbade any breaking or tearing and allowed only bending, stretching, and squeezing?"[41]

This passage helps to explain my portrayal of the geography of digestion. Latour offers a language to start integrating what geographers might refer to as a set of different scales. Starting at a local site—the gastrointestinal tract, in this case—and laying connections to other places in order to uncover the meaning of that local site is very much the geographical project here. This book has been intentionally organized into a set of four scales—body, building, city, hinterland—not to keep them separated, but in order to show how they demand integration. A helpful distinction is made here between a category of practice and one of analysis.[42] For the sake of an entry point, I adopt the practical scalar categories of body, building, and so on, but through the analysis we should arrive at a take-home message that teaches us that bodies and organs cannot be thought of as bound, isolated, and nested, but rather can be read from the landscapes that surround us.

Latour warns that thinking of a local site as inside larger frameworks—what he calls context, and what others have referred to as nested scales—forces a researcher to "jump" from one unit of analysis to another. The way to theoretically erase the breaks, tears, or jumps in scale is to instead bend, stretch, and squeeze—or, as I would have it, *extend*—the elements of a practical scale into one another. With digestion this means that we have looked at the origins of food, the technologies and ideologies related to its production, but also the excrement of human food waste. Where does it go after the body is done with it? What are the technologies and ideologies wrapped up in that process? To extend the food chain as it returns back into the city, the river, and the land is to consider the role of the place of transformation—the gastrointestinal tract. Understanding the digestive tract—and again, the technologies and ideologies connected to it—plays a crucial role in bending the practical categories that are so easy to keep intellectually separated.

If you were to remember something from this book—tomorrow as you're eating your corn flakes—I would hope it might be this: If we start at the gastrointestinal tract and trace outward, following the pathways of the objects and institutions related to this place—a place that historical sociologist Melanie DuPuis has called "Enteria"[43]—we can map the boundaries of the body, showing us that it is actually a range of places, practices, and beliefs. Our bodies are tied with the outside world in often unrevealed ways, and paying attention to those connections opens creative pathways to represent how bodies and environments are co-constituted. With the sustained interest in food studies and health geography, I hope to have contributed a methodology that offers a way to spatialize bodies so as to create a deeper understanding of the geographical consequences of human health and eating.

Landscape implies a certain spatial scale, though not one that can be simply measured. Instead, its size depends on its relation to the subject—that is, the person viewing it or moving through it. A landscape must, in my view, include a number of distinct places, objects, and people, most of which must be visible from a vantage point.[44] Its size, therefore, suggests an accessibility, an ability to move among the places that constitute it. Experientially, a landscape is between being there—that is, direct sensorial experience—and looking at a map of the area—that is, complete abstract detachment. It is capturing distant places with the eye, and knowing that those places are attainable with some effort. It is understanding one's surroundings without being physically present in all of it.

Why is landscape important for the geography of digestion? Because, in a general sense, landscape depends in various ways on bodies. There are some ready observations, such as that it takes (certain) bodies to construct buildings, or that cities cannot be cities without people's bodies inside them. But there are other insights, too. For example, the size of a landscape is figured with respect to the subject who is experiencing it. It is related to what can be captured visually from a very human, optical vantage point. A landscape is not initially, therefore, that which is directly experienced, but that which is first cognized with an eye, *then* traversed by a body. It is this moment in between seeing and "moving through" where we find the concept landscape.[45] Landscape directly precedes locomotion, action, and movement. It is, in this sense, completely dependent on the body position of the person doing the seeing, and the kinesthetic capacity of a body to be inside the territory captured by vision.[46] Landscape is the spatial expression of the subject/object duality, or rather, the spatial attempt to dissolve that duality. There can be no landscape for a body without a mind to cognize it, and there can be no landscape for a mind without a body to move through and touch it.

The insight that landscape is the spatial attempt to dissolve subject/object duality drives the work done in this book about digesting food in Battle Creek.[47] But we need to take this in a slightly different direction for it to help make sense of the geography of digestion in a more specific way. The concept of landscape merges our capacity to see and imagine territory with our capacity to go there and touch it.[48] Seeing from a distance suggests a connection between the subject doing the seeing and the objects being seen. The theoretical leap I am making moves us from the connectivity between the *eye* and the land to the connectivity between the *body* and the land. I am trying to clearly describe the coexistence of objects in a landscape with the object of the body. Landscape, as it is the spatial solution for dissolving subject/object, is as much the spatial solution for dissolving object uniquity. There is a spatial coincidence—an ontology, really—that I have sought to describe among bodies and the built environment. This ontology is not based on a body moving through a landscape with a head attached, interacting with material, touching it, altering it, then moving on. Instead it is based on seeing landscapes as extensions of our material bodies, about seeing ourselves in the very scene we capture with our vision. It is about being in the landscape without having to move through it. It is about being here and there at the same time. In this geography of digestion a body is *of* a landscape, not *in* it. We see a dissolution of the boundary between biology and artifice. When technologies are a part of the body, it encourages us to think about how we define the word *body* in the first place. Why should it be restricted to biological processes? A body is live cells, dead cells, many minerals, all constantly in flux.[49] For bodies surrounded by a built environment, part of this flux extends to those objects—as one urban geographer put it, the built environment is an "exoskeleton" of our bodies.[50] Pieces of our bodies need not be contained by our skin for them to be a legitimate part of our functioning, nor do we even necessarily need to be conscious of them.

Epilogue

Considering the rich association that John Kellogg maintained with the nineteenth-century health reform movement, including the rejection of pharmaceutical advances, it may be surprising that he remained committed to advancing his own practice through the use of chemical analysis and food-manufacturing technologies. Believing that digestion would solve all matters of health, the manipulation of foods to harness the curative properties of the digestive system was, for Kellogg, the next logical step in modernizing the eating philosophy that Sylvester Graham started.

Looking closely at Kellogg's practice at the Battle Creek Sanitarium, the contours of a lasting cultural conversation about food begin to emerge. That conversation is based on the search for foods that can harm us, and the subsequent reconfiguring of foods with technologies such that they are no longer detrimental. I believe this conversation has been the driver for much of the culture of food consumption in the United States through the twentieth century and up to today. Kellogg introduced a notion about food that might be described as "nature-plus," where the "plus" is a technological remediation—a comforting, modern fix that guarantees a certain state of corporeal health that nature itself cannot promise. That "alternative" food movements—organic, local, natural, nonindustrial—have made such strides over the past forty years in popular discourse clearly indicates the very baseline structure that they are up against: namely, a food system in which the value of a food has come to be determined by how it is augmented from a supposed state of nature. Pasteurization, additives, supplements, preservatives, and packaging (to name a few) are all present-day technological interventions that are meant to make food safer.

Distributing flaked cereal at the national scale fostered, and responded to, a consumer base that trusted science more than nature. It was a consumer base to whom the Kelloggs' Waxtite packaging conclusively indicated that the contents therein had been deemed exactly ready to eat by people who knew more about consumers' bodies than did the consumers doing the eating. Despite strong sub-economies today that foster direct farmer–consumer relationships, this deep trust in the science of food safety remains strong in American eating culture.

There are no doubt other modern stories in which technologies ameliorate the deleterious effects of bad food—milk being one of the main ones. With cereal, though, we see the contours of yet another lasting conversation about food begin to emerge. The recent ancestor of what we today recognize as breakfast cereal was engineered to most efficiently fit a particular set of body organs—the digestive system. Debates about digestive health are stronger than ever in national conversations about food and diet, with gluten intolerance, celiac disease, irritable bowel syndrome, and Crohn's disease being some of the most prominently recognized digestive disorders. There is also the probiotics movement, which promises to treat digestive disorders ranging from diarrhea to intestinal infections to irritable bowel syndrome.

With probiotics especially, the conversation is ripe to think about the digestive system as its own ecosystem. Ingesting probiotics with meals, a patient consumes bacteria that promote a certain state—a healthy state—of relationships among microorganisms in the gut such that the incorporation of food nutrients into the body happens efficiently and seamlessly. The ecosystem metaphor is powerful in the probiotics movement because it aligns with the same environmentalist discourse that blames industrial agriculture for flattening biological diversity where crops are grown. Encouraging biodiversity in the gut is as much a political move, then, as it is a health move. Consuming probiotics is a tacit statement that, at the very least, one can save the environment of his or her insides, even as environmental degradation occurs at dystopian proportions all around.

It is exactly this connection between inside and outside that has been explored in *A Geography of Digestion.* John Kellogg certainly thought that the digestive system was its own ecosystem, too. And though he thought the gut's ecosystem should be sterile to achieve well-being (the opposite of probiotics), he made the connection to achieving that state with the outside world. By directly influencing the built landscapes surrounding Battle Creek, he mirrored his vision of a clean, efficient stomach in the clean, efficient landscapes of food manufacturing, sewage removal, and agricultural production. Kellogg needed the surrounding landscapes to make his modern stomachs at the sanitarium, just as progressive food movements today need local landscapes to do work for their bodies by making fresh, organic, or unadulterated food. Kellogg's vision was, in short, a geographical one.

Kellogg's, then, is a prime case in which we can interpret, or read, the digestive system from the surrounding landscapes. This practice need not stop with Kellogg at the end of the nineteenth century, though. What do landscapes of irritable bowel syndrome look like? Or landscapes of gluten intolerance? Though it may not seem common to think of these digestive conditions in such terms, I believe there is already a language in place to achieve this greater geographical awareness. Picturing the unending fields of corn in the upper Midwest, for example, or in California's smog-soaked Central Valley, might help us comprehend a greater, national agriculture food system, its distribution channels, and its ultimate endpoint—our very own stomachs—every time we eat. *A Geography of Digestion* reminds us that we are much more than what we eat; we are the places we make and inhabit.

Notes

INTRODUCTION

1. Stacy Alaimo, "Thinking as the Stuff of the World," *O-Zone: A Journal of Object-Oriented Studies* 1 (2014): 14.

2. The racial politics of eating, using digestion as a metaphor and as a material process, are explored in E. Melanie DuPuis, *Dangerous Digestion: The Politics of American Dietary Advice* (Oakland: University of California Press, 2015).

3. John H. Kellogg, *The Stomach: Its Disorders and How to Cure Them* (Battle Creek, Mich.: Modern Medicine Publishing Co., 1896), 3.

4. National Digestive Diseases Information Clearinghouse, *Irritable Bowel Syndrome* (Bethesda, Md.: National Institute of Diabetes and Digestive and Kidney Diseases, 2013).

5. Melissa L. Caldwell, "Digestive Politics in Russia: Feeling the Sensorium beyond the Palate," *Food and Foodways* 22, nos. 1–2 (2014); and DuPuis, *Dangerous Digestion.*

6. For two examples, see Nancy Langston, *Toxic Bodies: Hormone Disruptors and the Legacy of DES* (New Haven: Yale University Press, 2010); and Norah Anita Schwartz et al., "'Where They (Live, Work and) Spray': Pesticide Exposure, Childhood Asthma and Environmental Justice among Mexican-American Farmworkers," *Health and Place* 32 (2015).

7. Jack Ralph Kloppenburg, *First the Seed: The Political Economy of Plant Biotechnology, 1492–2000* (New York: Cambridge University Press, 1988);

Daniel Buck, Christina Getz, and Julie Guthman, "From Farm to Table: The Organic Vegetable Commodity Chain of Northern California," *Sociologia Ruralis* 37, no. 1 (1997).

8. See also Gregg Mitman, "In Search of Health: Landscape and Disease in American Environmental History," *Environmental History* 10, no. 2 (2005): 104–210.

9. The topic of "how consumption begins with changes to the material world, to physical nature" has been taken up by, among others, Matthew W. Klingle, "Spaces of Consumption in Environmental History" *History and Theory* 42, no. 4 (2003): 94–110.

10. A recent geographical take that addresses actor network theory is found in Scott Kirsch, "Cultural Geography III: Objects of Culture and Humanity, or, Re-'Thinging' the Anthropocene Landscape," *Progress in Human Geography* 39, no. 6 (2015). For an introduction to object-oriented ontology, see Graham Harman, *Guerrilla Metaphysics: Phenomenology and the Carpentry of Things* (Chicago: Open Court, 2005).

11. Much thought in cultural geography has been given to the practice of description. For example, see David Matless, "Describing Landscape: Regional Sites," *Performance Research* 15, no. 4 (2010). The roots of description in American geography are often traced to the regional geography of Richard Hartshorne. See Richard Hartshorne, *The Nature of Geography: A Critical Survey of Current Thought in the Light of the Past* (Lancaster, Pa.: The Association of American Geographers, 1939).

12. Foucault's concept of biopolitics has spawned a vast literature unto itself. One of Foucault's original articulations is Michel Foucault, *The History of Sexuality, Volume I: An Introduction,* trans. R. Hurley (New York: Vintage Books, 1990; originally published in 1976). While my treatment is not strictly Foucauldian, there is no doubt that another book could be written under such theoretical concerns.

13. See also Thomas Osborne, "Security and Vitality: Drains, Liberalism and Power in the Nineteenth Century," in *Foucault and Political Reason: Liberalism, Neo-Liberalism and Rationalities of Government,* ed. A. Barry, T. Osborne, and N. Rose (Chicago: University of Chicago Press, 1996).

14. There is a significant literature on John Kellogg concerning his research, thinking, and practice on issues other than eating and digestion throughout his career, including eugenics, sexual abstinence, and masturbation. Key texts here include: Ronald Numbers, "Sex, Science, and Salvation: The Sexual Advice of Ellen G. White and John H. Kellogg," in *Right Living: An Anglo-American Tradition of Self-Help Medicine and Hygiene,* ed. C.E. Rosenberg (Baltimore, Md.: The Johns Hopkins University Press, 2003); John Money, *The Destroying Angel: Sex, Fitness and Food in the Legacy of Degeneracy Theory, Graham Crackers, Kellogg's Corn Flakes and American Health History* (Buffalo, N.Y.: Prometheus

Books, 1985); and Stephen Nissenbaum, *Sex, Diet, and Debility in Jacksonian America: Sylvester Graham and Health Reform* (Westport, Conn.: Greenwood Press, 1980). I have chosen for this book to focus on Kellogg's philosophy of eating and digestion.

15. Atwater will resurface in chapters 2 and 5. For an introduction, see Kenneth Carpenter, "The Life and Times of W. O. Atwater (1844–1907)," *Journal of Nutrition* 124 (1994).

16. For a recent and thorough take on Kellogg's medical philosophies vis-à-vis his crucial affiliation with Seventh-day Adventism, see Brian C. Wilson, *Dr. John Harvey Kellogg and the Religion of Biologic Living* (Bloomington: Indiana University Press, 2014).

17. However, what the food industry made in the mid-twentieth century was not universally adopted by American housewives. See Laura Shapiro, *Something from the Oven: Reinventing Dinner in 1950s America* (New York: Viking, 2004); and David Strauss, *Setting the Table for Julia Child: Gourmet Dining in America, 1934–1961* (Baltimore: Johns Hopkins University Press, 2011).

18. Kelly Dowhower Karpa, *Bacteria for Breakfast: Probiotics for Good Health* (Victoria, B.C.: Trafford, 2003).

19. "101 and We've Just Begun: Cereal Manufacturing Companies in Battle Creek," *Heritage Battle Creek: A Journal of Local History* 2 (Spring 1992).

20. With the perspective that technicians introduce safety to food, it is fitting that cereal and milk should have grown up together. For the story on milk's rise, see E. Melanie DuPuis, *Nature's Perfect Food: How Milk Became America's Drink* (New York: New York University Press, 2002).

21. Advertisement, *Modern Medicine and Bacteriological Review* 3, no. 2 (February 1894).

22. Martin V. Melosi, *The Sanitary City: Urban Infrastructure in America from Colonial Times to the Present* (Baltimore: Johns Hopkins University Press, 2000).

23. In addition to medical concerns, French historian Alain Corbin has shown some of the cultural elements of nineteenth-century modernism. For his examination on the role of smell, see Alain Corbin, *The Foul and the Fragrant: Odor and the French Social Imagination* (Cambridge, Mass.: Harvard University Press, 1986).

24. N. Katherine Hayles, *How We Became Posthuman: Virtual Bodies in Cybernetics, Literature, and Informatics* (Chicago: University of Chicago Press, 1999), 2–3.

25. Though for a move in this direction, see the 2006 special issue of *Social and Cultural Geography*, called "Posthuman Geographies." In particular, see Noel Castree and Catherine Nash, "Editorial: Posthuman Geographies," *Social and Cultural Geography* 7, no. 4 (2006); and Heike Jons, "Dynamic Hybrids and the Geographies of Technoscience: Discussing Conceptual Resources Beyond the Human/Non-Human Binary," *Social and Cultural Geography* 7, no. 4 (2006).

26. Diana Coole and Samantha Frost, "Introducing the New Materialisms," in *New Materialisms: Ontology, Agency, and Politics,* ed. D. Coole and S. Frost (Durham, N.C.: Duke University Press, 2010), 3.

27. For more about how this paradox between nature and technology played out with respect to milk, see Kendra Smith-Howard, *Pure and Modern Milk: An Environmental History since 1900* (New York: Oxford University Press, 2014).

28. A foundational book on the 1960s counterculture food movements that renounced technologically modified foods (among other things) is Warren J. Belasco, *Appetite for Change: How the Counterculture Took on the Food Industry,* 2nd ed. (Ithaca: Cornell University Press, 2007). The leading take on the cultural politics of dietary reform in America is Charlotte Biltekoff, *Eating Right in America: The Cultural Politics of Food and Health* (Durham: Duke University Press, 2013).

1. THE BATTLE CREEK SANITARIUM

1. Roy E. Graham, *Ellen G. White: Co-Founder of the Seventh-day Adventist Church* (New York: Peter Lang, 1985), 70; and Ronald L. Numbers, *Prophetess of Health: Ellen G. White and the Origins of Seventh-day Adventist Health Reform* (Knoxville: University of Tennessee Press, 1992), xiii–xiv.

2. Malcolm Bull and Keith Lockhart, *Seeking Sanctuary: Seventh-day Adventism and the American Dream* (Bloomington: Indiana University Press, 2007), 128.

3. Graham, *Ellen G. White,* 82–83.

4. Richard William Schwarz, "John Harvey Kellogg: American Health Reformer" (dissertation, History, University of Michigan, Ann Arbor, 1965), 176.

5. Ronan Foley, *Healing Waters: Therapeutic Landscapes in Historic and Contemporary Ireland* (Burlington, Vt.: Ashgate, 2010), 54.

6. Wilbert M. Gesler, *Healing Places* (New York: Rowman and Littlefield, 2003). Here Gesler uses the examples of Epidauros, Greece; Bath, England; and Lourdes, France.

7. David Livingstone, *Putting Science in Its Place: Geographies of Scientific Knowledge* (Chicago: University of Chicago Press, 2003).

8. Alan R. Beebe, "History of the Battle Creek Sanitarium, 1866–1903" (unpublished thesis, Kalamazoo College, held at the Willard Library, Helen Warner Branch, Battle Creek, Mich., 1949).

9. Numbers, *Prophetess of Health,* 105–107; James Whorton, *Crusaders for Fitness: The History of American Health Reformers* (Princeton, N.J.: Princeton University Press, 1982), chapter 2.

10. Beebe, "History of the Battle Creek Sanitarium."

11. Numbers, *Prophetess of Health,* 105 (emphasis in original).

12. For examples of the myriad statements of health from the Western Health Reform Institute, see their journal *The Health Reformer*. These examples are from an 1868 issue (vol. 3, no. 2).

13. "A Wisconsin Opinion of the Sanitarium," *The Daily Inter Ocean*, ca. 1890. Housed at the Bentley Historical Library, University of Michigan.

14. Scott Bruce and Bill Crawford, *Cerealizing America: The Unsweetened Story of American Breakfast Cereal* (Boston: Faber and Faber, 1995), 7.

15. Greater context on the relationship between bodies and environments in the nineteenth-century United States is given in Conevery Bolton Valencius, *The Health of the Country: How American Settlers Understood Themselves and Their Land* (New York: Perseus Books, 2002). See especially chapter 2, "The Body." And Linda Nash, *Inescapable Ecologies: A History of Environment, Disease, and Knowledge* (Berkeley: University of California Press, 2006).

16. James C. Jackson, *How to Treat the Sick without Medicine* (New York: Austin, Jackson and Co., 1871), 18–25.

17. Ibid., 25.

18. Harry B. Weiss and Howard R. Kemble, *The Great American Water-Cure Craze: A History of Hydropathy in the United States* (Trenton, N.J.: The Past Times Press, 1967), 41.

19. When Priessnitz moved to Jesenik the town was called Graefenberg. This is not to be confused with the present-day Graefenberg, Germany, outside of Nuremberg.

20. Richard Metcalfe, *Life of Vincent Priessnitz: Founder of Hydrotherapy* (London: Simpkin, Marshall, Hamilton, Kent and Co., 1898), 133.

21. Weiss and Kemble, *The Great American Water-Cure Craze*, 4–5. Also of note, the Eastern European health retreat genre is used as a site of fiction in Wes Anderson's film *The Grand Budapest Hotel* (United States: Fox Searchlight, 2014).

22. Historian of food and technology Carolyn Thomas (formerly de la Peña) has written about the development of water cures in the United States further into the twentieth century. She also focuses on the implementation of machinery, discussing many of the same machines Kellogg used at the Battle Creek Sanitarium. See Carolyn de la Peña, "Recharging at the Fordyce: Confronting the Machine and Nature in the Modern Bath," *Technology and Culture* 40, no. 4 (1999): 746–769.

23. As of 2014 the Priessnitz Medical Spa operates at the original site of Priessnitz's water cure institute, utilizing his legacy to sell healthful vacationing. See http://en.priessnitz.cz/.

24. Susan E. Cayleff, *Wash and Be Healed: The Water-Cure Movement and Women's Health* (Philadelphia: Temple University Press, 1987), 19.

25. Ronald Numbers, "Sex, Science, and Salvation: The Sexual Advice of Ellen G. White and John H. Kellogg," in *Right Living: An Anglo-American Tradition*

of Self-Help Medicine and Hygiene, ed. C.E. Rosenberg (Baltimore, Md.: The Johns Hopkins University Press, 2003).

26. Numbers, *Prophetess of Health*, 83–84.

27. Marshall Scott Legan, "Hydropathy in America: A Nineteenth Century Panacea," *Bulletin of the History of Medicine* 45, no. 3 (1971): 272.

28. Schwarz, "John Harvey Kellogg," 8–10.

29. Numbers, *Prophetess of Health*, 117.

30. Schwarz, "John Harvey Kellogg," 173.

31. J.N. Loughborough, "Report from Bro. Loughborough," *Review and Herald* 28 (September 1866), 117. From Numbers, *Prophetess of Health*, 105.

32. Schwarz, "John Harvey Kellogg," 20–30.

33. Numbers, *Prophetess of Health*, 126.

34. *Vis medicatrix Naturae* ("the healing power of nature") is from Hippocrates. The quotes are from John H. Kellogg, "The Ideal Sanitarium" (lecture read before the American Climatological Society, September 2, 1890), 2–3.

35. Ibid., 4.

36. See also Kendra Smith-Howard, *Pure and Modern Milk: An Environmental History since 1900* (New York: Oxford University Press, 2014); and Gregg Mitman, *Breathing Space: How Allergies Shape Our Lives and Landscapes* (New Haven: Yale University Press, 2007).

37. "A Health Resort," *The Sunday Inter Ocean* 23, no. 134 (1894).

38. R.T. Claridge, *Hydropathy; Or, the Cold Water Cure, as Practiced by Vincent Priessnitz, at Graefenberg, Silesia, Austria*, 8th ed. (London: James Madden and Co., 1843), 14.

39. Jennifer Lea, "Retreating to Nature: Rethinking 'Therapeutic Landscapes,'" *Area* 40, no. 1 (2008): 91.

40. In 1844, Englishman Robert Hay Graham wrote a report on Priessnitz, entitled *Graefenberg; Or, a true Report of the Water-Cure, with an Account of its Antiquity*, in which he discredits the originality of the institution and states that only about one in twenty patients was "cured." See Weiss and Kemble, *The Great American Water-Cure Craze*, 12–13.

41. Definitions of *sanatorium* and *sanitarium* are from the online edition of the *Oxford English Dictionary*.

42. Bruce and Crawford, *Cerealizing America*, 14.

43. Ronan Foley, "Performing Health in Place: The Holy Well as a Therapeutic Assemblage," *Health and Place* 17, no. 2 (2011).

44. Robert David Sack, *Homo Geographicus: A Framework for Action, Awareness, and Moral Concern* (Baltimore, Md.: Johns Hopkins University Press, 1997).

45. Kellogg, "The Ideal Sanitarium," 5.

46. The first of Kellogg's divergences is pointed out in Bull and Lockhart, *Seeking Sanctuary*, 130. The second and third are my claims.

47. Ellen White, *Health, or, How to Live* (Battle Creek, Mich.: Steam Press of the Seventh-day Adventist Publishing Association, 1865), 8–20. Available at http://text.egwwritings.org/.

48. John H. Kellogg, *The Uses of Water in Health and Disease: A Practical Treatise on the Bath, Its History and Uses* (Battle Creek, Mich.: The Health Reformer, 1876), 75.

49. The novel that has spawned the most recent installment of cultural amusement with the Battle Creek Sanitarium is T. Coraghessan Boyle, *The Road to Wellville* (New York: Penguin, 1993). From this came the 1994 film *The Road to Wellville* (dir. Alan Parker), starring Anthony Hopkins, Bridget Fonda, and Matthew Broderick. More recently still, the 2013 episode "Detroit" in Comedy Central's television series *Drunk History* (dir. Jeremy Konner), starring Owen Wilson and Luke Wilson, is about John Kellogg and his health practices at the Sanitarium.

50. Claridge, *Hydropathy*, 152, 273.

51. Jackson, *How to Treat the Sick without Medicine*, 32–34.

52. Smith-Howard, *Pure and Modern Milk*.

53. For more on this obsession as it relates to obesity, see Julie Guthman, *Weighing In: Obesity, Food Justice, and the Limits of Capitalism* (Berkeley: University of California Press, 2011).

54. I have not yet seen any evidence of Native Americans, African Americans, or immigrants from any point of origin who stayed at the Battle Creek Sanitarium. The tacit cultural knowledge of how to behave at a water cure was, apparently and not surprisingly, deeply racialized.

55. Advertisement found in *The Bacteriological World and Modern Medicine* 1, no. 11 (September 1892): 385–386.

56. John H. Kellogg, "Sanitarium Lectures: Nature's Method of Defending the Body against Disease," June 5, 1891 (manuscript held at the Bentley Historical Library, University of Michigan), 4–5.

57. Vincent Priessnitz, *The Cold Water Cure: Its Principles, Theory, and Practice with Ample Directions for Its Self-Application* (London: William Strange, 1843).

2. SCIENTIFIC EATING

1. On John Kellogg's request, his wife, Ella Eaton Kellogg, initiated an experiment kitchen in the Sanitarium in 1883 and directed its operations for over twenty years.

2. John H. Kellogg, *The Stomach: Its Disorders and How to Cure Them* (Battle Creek, Mich.: Modern Medicine Publishing Co., 1896), 3, 135.

3. John H. Kellogg, "The New Chemistry of the Stomach," *The Bacteriological World and Modern Medicine* 1, no. 13 (1892): 433.

4. John Kellogg was known as an accomplished gastroenterological surgeon and was said to have performed thousands of abdominal surgeries without a death. See La Scienya Jackson, Stanley Dudrick, and Bauer Sumpio, "John Harvey Kellogg: Surgeon, Inventor, Nutritionist (1852–1943)," *Journal of the American College of Surgeons* 199, no. 5 (2004). Kellogg published widely on his surgical techniques. See, e.g., John H. Kellogg, "A New Procedure for Suspension of the Pelvic Colon," *The Bulletin of the Battle Creek Sanitarium and Hospital Clinic* 18, no. 2 (1922). Alarmingly, Richard Schwarz reports that by the end of his life Kellogg had performed over 22,000 operations. See Richard Schwarz, "John Harvey Kellogg: American Health Reformer" (dissertation, History, University of Michigan, Ann Arbor, 1965), 275.

5. Charles E. Rosenberg, *The Care of Strangers: The Rise of America's Hospital System* (New York: Basic Books, 1987), 169–173. In the 1870s and 1880s this was rapidly changing, though, toward specialization. Kellogg practiced as both a generalist and a specialist.

6. Joseph B. Kirsner, *The Development of American Gastroenterology* (New York: Raven Press, 1990), 96.

7. "About Battle Creek," *Headlight: A Periodical Devoted to the Interests of Railroads and Railroad Centers,* May 1895, 22.

8. S. M. Baker, "Nursing in Homes, Private Hospitals, and Sanitariums," *Modern Medicine and Bacteriological Review* 3, no. 4 (1894).

9. "A Health Resort," *The Sunday Inter Ocean* 23, no. 134 (1894). The *Daily Inter Ocean* was a Chicago-based newspaper, issues of which are housed at the Bentley Historical Library, University of Michigan.

10. Patsy Gerstner, "The Temple of Health: A Pictorial History of The Battle Creek Sanitarium," *Caduceus* 12, no. 2 (1996), 42.

11. See "A Health Resort." And *Calhoun County Souvenir: Commemorating the Battle Creek Journal's Fiftieth Anniversary: Calhoun County, Her Industries and Sketches of Her Representative Citizens* (Battle Creek, Mich.: Battle Creek Journal, 1901), 79.

12. Although chemists now would not measure for chlorine and protein, they did at the time.

13. John H. Kellogg, "Cancer of the Stomach" (paper read before the Michigan State Medical Society, 1897), 1–11.

14. In 1822, under the employ of the American Fur Company, a French-Canadian named Alexis St. Martin was accidentally shot at nearly point-blank range while working in northern Michigan. The "powder and duck shot" blasted open a hole into St. Martin's stomach, but miraculously did not kill him. Interested in the physiology of digestion, and recognizing the rare opportunity that St. Martin's newly opened stomach afforded, the local physician William Beaumont followed and at times enslaved St. Martin for the next eleven years as his object of study. In this case, peering into the digestive system was more obtrusive

than was Kellogg's chemical method. See William Beaumont, *Experiments and Observations on the Gastric Juice, and the Physiology of Digestion* (Plattsburgh, N.Y.: F. P. Allen, 1833).

15. Kellogg, *The Stomach,* 130.

16. "They are inclined, from curiosity to learn what progress they are making. . . . Patients are always deeply interested in the results of the chemical investigation of the stomach fluid." From Kellogg, "The New Chemistry of the Stomach," 434.

17. Kellogg, *The Stomach,* 143.

18. Nancy Tomes, *The Gospel of Germs: Men, Women, and the Microbe in American Life* (Cambridge, Mass.: Harvard University Press, 1998), 39.

19. Kellogg, *The Stomach,* 139.

20. John H. Kellogg, "Decomposing Organic Matter" (paper read before the Sanitary Convention, Michigan State Board of Health, 1882), 226.

21. Robert P. Hudson, "Theory and Therapy: Ptosis, Stasis, and Autointoxication," *Bulletin of the History of Medicine* 63, no. 3 (1989): 396–397.

22. Harvey Diamond and Marilyn Diamond, *Fit for Life* (New York: Warner Books, 1985), 6, 48.

23. Charles Bouchard, *Lectures on Auto-Intoxication in Disease; Or, Self-Poisoning of the Individual* (Philadelphia: F. A. Davis, 1894), 5. Originally published in 1887.

24. Hudson, "Theory and Therapy," 397.

25. Kellogg, *The Stomach,* 160.

26. Ibid., 161.

27. Ibid., 353.

28. Brian C. Wilson, *Dr. John Harvey Kellogg and the Religion of Biologic Living* (Bloomington: Indiana University Press, 2014), 49.

29. A foundational work on the history of medical typologies is Geoffrey Bowker and Susan Leigh Star, *Sorting Things Out: Classification and Its Consequences* (Cambridge, Mass.: MIT Press, 1999).

30. Kellogg, "The New Chemistry of the Stomach," 434.

31. Jacqueline Urla and Jennifer Terry, "Introduction: Mapping Embodied Deviance," in *Deviant Bodies: Critical Perspectives on Difference in Science and Popular Culture,* ed. J. Terry and J. Urla (Bloomington: Indiana University Press, 1995), 1–11. See also E. Melanie DuPuis, *Dangerous Digestion: The Politics of American Dietary Advice* (Oakland: University of California Press, 2015).

32. John H. Kellogg, "The Ideal Sanitarium" (lecture read before the American Climatological Society, September 2, 1890), 2.

33. Michael McGerr, *A Fierce Discontent: The Rise and Fall of the Progressive Movement in America, 1870–1920* (New York: Free Press, 2003).

34. It would be easy to assume that John Kellogg intentionally manipulated his operations to squeeze profits from unsuspecting patients. There is little to

nothing, however, in his hundreds of nineteenth-century writings that indicates anything but earnest sincerity about trying to build human health. This would change in the early twentieth century as the business relationship he had with his brother, William Keith Kellogg, intensified and became a power struggle for assets.

35. John H. Kellogg, "Relation of Modern Physiological Chemistry to Vegetarianism," *Modern Medicine and Bacteriological Review* 3, no. 3 (1894): 58.

36. Ibid., 83–84.

37. John H. Kellogg, "The Influence of an Aseptic Dietary upon the Bacteria of the Stomach," *Modern Medicine and Bacteriological Review* 4, no. 8 (1895): 217.

38. "Michigan Weather and Enterprise," *The Daily Inter Ocean,* ca. 1880. Housed at the Bentley Historical Library, University of Michigan.

39. Kellogg, "The Ideal Sanitarium," 11.

40. *Calhoun County Souvenir,* 83.

41. John H. Kellogg, *The Battle Creek Sanitarium System: History, Organization, Methods* (Battle Creek, Mich.: Gage Printing Co., 1908), 119.

42. S. M. Baker, "Nursing in Homes, Private Hospitals, and Sanitariums," *Modern Medicine and Bacteriological Review* 3, no. 4 (1894): 77–82.

43. Through the 1880s and 1890s Kellogg's practice helped him develop what would by the late 1920s become a more nuanced stance on the importance of intestinal flora. Throughout his career, though, he maintained an obsession with fostering the conditions for high quantities of acid in the stomach. For examples, see John H. Kellogg, "Dependable Methods of Producing and Maintaining an Aciduric Intestinal Flora," reprinted from the *Medical Journal and Record,* September 18, 1929; and John H. Kellogg, "A Successful and Dependable Method of Changing the Intestial Flora," *Bulletin of the Battle Creek Sanitarium and Hospital Clinic* 22 (1928): 115.

44. Kellogg, "The New Chemistry of the Stomach," 434.

45. Sarah Stage, "Ellen Richards and the Social Significance of the Home Economics Movement," in *Rethinking Home Economics: Women and the History of a Profession,* ed. S. Stage and V. B. Vincenti (Ithaca: Cornell University Press, 1997), 23.

46. Ella E. Kellogg, *Science in the Kitchen: The Principles of Healthful Cookery* (Battle Creek, Mich.: Modern Medicine Publishing Co., 1893), 3.

47. Ibid., 60–61.

48. Foucault's ideas are from his book *Discipline and Punish: The Birth of the Prison,* trans. A Sheridan (New York: Vintage Books, 1995; originally published in 1975). I took the quote from Paul Rabinow, ed., *The Foucault Reader* (New York: Pantheon Books, 1984), 190.

49. Nancy Tomes, "Spreading the Germ Theory: Sanitary Science and Home Economics, 1880–1930," in *Rethinking Home Economics: Women and the History of a Profession,* ed. S. Stage and V.B. Vincenti (Ithaca: Cornell University Press, 1997), 34.

50. Note from John Kellogg to William Kellogg (typed manuscript, September 28, 1898), Michigan State University Archives and Historical Collections, John H. Kellogg Papers, East Lansing, Mich.

51. Kellogg, *The Stomach,* 223.

52. Kellogg, *The Stomach,* 224.

53. Kellogg, *The Stomach,* 230.

54. John H. Kellogg, *Balanced Bills of Fare: Arranged with Reference to the Normal Daily Ration; And the Needs of Special Classes of Invalids* (Battle Creek, Mich.: Good Health Pub. Co., 1899).

55. John H. Kellogg, "Sanitarium Lectures: How to Live a Century," 1884 (manuscript held at the Bentley Historical Library, University of Michigan).

56. Scott Bruce and Bill Crawford, *Cerealizing America: The Unsweetened Story of American Breakfast Cereal* (Boston: Faber and Faber, 1995), 20.

57. Horace B. Powell, *The Original Has This Signature—W.K. Kellogg: The Story of a Pioneer in Industry and Philanthropy* (Battle Creek, Mich.: W.K. Kellogg Foundation, 1956), 87.

58. "Dr. Kellogg Recalls Early Sanitarium Diet Experiments," *Battle Creek Moon-Journal* 24 (November 21, 1938). See also Gerald Carson, *Cornflake Crusade* (New York: Rinehart and Co., 1957), 117.

59. Kellogg, *Balanced Bills of Fare,* 1–2.

60. E. Kellogg, *Science in the Kitchen,* 3.

61. Justus von Liebig, *Organic Chemistry in its Applications to Agriculture and Physiology* (London: Taylor and Walton, 1840).

62. Margaret W. Rossiter, *The Emergence of Agricultural Science: Justus Liebig and the Americans, 1840–1880* (New Haven: Yale University Press, 1975).

63. Justus von Liebig, *Familiar Letters on Chemistry, and its Relation to Commerce, Physiology and Agriculture* (Philadelphia: Campbell, 1843).

64. Edward H. Beardsley, *The Rise of the American Chemistry Profession, 1850–1900* (Gainesville: University of Florida Press, 1964), 17.

65. Ibid., 14.

66. Rossiter, *The Emergence of Agricultural Science,* 163.

67. Naomi Aronson, "Nutrition as a Social Problem: A Case Study of Entrepreneurial Strategy in Science," *Social Problems* 29, no. 5 (1982); Anson Rabinbach, *The Human Motor: Energy, Fatigue, and Origins of Modernity* (Berkeley: University of California Press, 1992).

68. Kenneth Carpenter, "The Life and Times of W.O. Atwater (1844–1907)," *Journal of Nutrition* 124 (1994).

69. Wilbur O. Atwater and Chas D. Woods, *The Chemical Composition of American Food Materials: Bulletin No. 28* (Washington, D.C.: Government Printing Office; U.S. Department of Agriculture Office of Experiment Stations, 1896).

70. L. J. Rather, "The 'Six Things Non-Natural': A Note on the Origins and Fate of a Doctrine and a Phrase," *Clio Medica* 3 (1968).

71. Kellogg, "The Influence of an Aseptic Dietary upon the Bacteria of the Stomach."

3. FLAKED CEREAL

1. For a popular example of the critique against industrially produced foods in the twenty-first century, see Michael Pollan, "Our National Eating Disorder," *New York Times Magazine*, October 17, 2004.

2. Richard William Schwarz, "John Harvey Kellogg: American Health Reformer" (dissertation, History, University of Michigan, Ann Arbor, 1965), 277.

3. Ella E. Kellogg, *Science in the Kitchen: The Principles of Healthful Cookery* (Battle Creek, Mich.: Modern Medicine Publishing Co., 1893), 91.

4. Gerald Carson, *Cornflake Crusade* (New York: Rinehart and Co., 1957), 67.

5. Heather Arndt Anderson, *Breakfast: A History* (Lanham, Md.: Rowman and Littlefield, 2013), 37.

6. Scott Bruce and Bill Crawford, *Cerealizing America: The Unsweetened Story of American Breakfast Cereal* (Boston: Faber and Faber, 1995), 8–21.

7. Schwarz, "John Harvey Kellogg," 278.

8. John H. Kellogg, "The Ideal Sanitarium" (lecture read before the American Climatological Society, September 2, 1890), 10.

9. Andrew F. Smith, *Eating History: 30 Turning Points in the Making of American Cuisine* (New York: Columbia University Press, 2009), 142.

10. Anderson, *Breakfast*, 37.

11. Sylvester Graham, *A Treatise on Bread, and Bread-Making* (Boston: Light and Stearns, 1837), 44.

12. Stephen Nissenbaum, *Sex, Diet, and Debility in Jacksonian America: Sylvester Graham and Health Reform* (Westport, Conn.: Greenwood Press, 1980), 4–5.

13. I say "perceived" local, social fabric here because in cities—especially in Europe—bread making was a profession already removed from the home by the 1830s.

14. James Whorton, *Crusaders for Fitness: The History of American Health Reformers* (Princeton, N.J.: Princeton University Press, 1982), 42.

15. Nissenbaum, in *Sex, Diet, and Debility in Jacksonian America*, offers a more in-depth study of how Graham wove together French medical theory to make sense of his own eating philosophies. See especially pp. 20–21.

16. Whorton, *Crusaders for Fitness*, 42–47.

17. The problem of tasteless, bland foods from the health reform movement would be entirely reversed in the early twentieth century by the addition of salts and sugars.

18. Schwarz, "John Harvey Kellogg," 419.

19. John H. Kellogg, *The Stomach: Its Disorders and How to Cure Them* (Battle Creek, Mich.: Modern Medicine Publishing Co., 1896), 44 (emphasis added).

20. John H. Kellogg, *The Battle Creek Sanitarium System: History, Organization, Methods* (Battle Creek, Mich.: Gage Printing Co., 1908), 137.

21. Thomas S. Kuhn, *The Structure of Scientific Revolutions* (Chicago: University of Chicago Press, 1962). The quote is from Derek Gregory, "Paradigm," in *The Dictionary of Human Geography*, ed. R. J. Johnston, D. Gregory, G. Pratt, and M. Watts (Malden, Mass.: Blackwell, 2000), 571.

22. In 1890 John Kellogg opened the Sanitas Food Company, which operated contemporaneously and somewhat separately from the already existing Battle Creek Sanitarium Health Food Company. The Sanitas Food Co. would first start selling the new, post-paradigm-shift line of products, including flaked cereals and vegetable meats, while the Sanitarium Health Food Co. maintained control over the production of granola and the older line of biscuits and breads. See Schwarz, "John Harvey Kellogg," 420; and Horace B. Powell, *The Original Has This Signature—W. K. Kellogg: The Story of a Pioneer in Industry and Philanthropy* (Battle Creek, Mich.: W. K. Kellogg Foundation, 1956), 89.

23. Schwarz, "John Harvey Kellogg," 419.

24. Carson, *Cornflake Crusade*, 122.

25. Richard Schwarz, *John Harvey Kellogg, M.D.: Pioneering Health Reformer* (Hagerstown, Md.: Review and Herald Pub. Association, 2006), 116; and Bruce and Crawford, *Cerealizing America*, 49.

26. Schwarz, *John Harvey Kellogg*, 116.

27. The details of who exactly performed specific actions at this point are contested for two important reasons. The first is a legal one. John Kellogg's primary biographer, Richard Schwarz, fashions much of his story from John's testimony in the Michigan State Supreme Court in 1920. The series of testimonies are the result of a major dispute between John and Will about who retained the right to market flaked cereals. The invention moment of flaked cereals, therefore, played a role in determining who controlled it as intellectual property. Interestingly, it is widely agreed among historians that John was trying to protect the sanctity and wholesomeness of corn and wheat flakes from being ushered into mass- and highly sweetened production. To John this would have divorced health foods from his holy health mission, but nonetheless this is exactly what Will desired in the name of growing the business. The second reason that the details of this lore are contested and blurred is because it is such a significant moment in the cultural history of Seventh-day Adventism. Because this food's origins play a part in

the collective identity of an entire religious establishment, it is a highly researched and described series of events with an extraordinary number of details that are employed differently by various storytellers. These storytellers are sometimes interested in the history of the church, but more often are interested in breakfast cereals for their kitschy American cultural appeal. There is Schwarz's story (which remains consistent in the 1965, 1970, and 2006 editions of his biography), the primary biography of Will Kellogg (Powell, *The Original Has This Signature*), and then the almost countless renditions that are derivations of one or both of these (prominent examples include Ronald M. Deutsch, *The Nuts among the Berries: An Exposé of America's Food Fads* [New York, Ballantine Books, 1967]; Bruce and Crawford, *Cerealizing America;* Smith, *Eating History;* and Anderson, *Breakfast*). Carson, in *Cornflake Crusade,* relies on primary documents and personal interviews, and the book is an early pillar in the historiography of the breakfast cereal industry. Of all these, Schwarz's 1965 dissertation provides the most robust and careful listing of primary documents, though the trustworthiness of those primary sources, too, are disputable. I have tried my best to triangulate the story to tell it as accurately as possible, mostly relying on Schwarz, Powell, and Carson.

28. Schwarz, *John Harvey Kellogg,* 116.

29. Powell, *The Original Has This Signature,* 91.

30. Robert B. Fast and Elwood F. Caldwell, *Breakfast Cereals and How They Are Made,* 2nd ed. (St. Paul, Minn.: American Association of Cereal Chemists, 2000). The quote is from the preview to chapter 5 in the online version, available at http://dx.doi.org/10.1094/1891127152.005.

31. Brian C. Wilson, *Dr. John Harvey Kellogg and the Religion of Biologic Living* (Bloomington: Indiana University Press, 2014).

32. John H. Kellogg, *Flaked Cereals and Process of Preparing Same* (United States: United States Patent Office, 1896).

33. Carson, *Cornflake Crusade,* 123–125.

34. Kellogg, *Flaked Cereals and Process of Preparing Same,* patent. All emphases added.

35. A popular articulation of this early twenty-first-century mistrust can be found in Michael Pollan, *The Omnivore's Dilemma: A Natural History of Four Meals* (New York: Penguin, 2006).

36. John H. Kellogg, "Experimental Research Relating to Salivary Secretion and Digestion," in *Report of the Laboratory of Hygiene connected with the Battle Creek Sanitarium,* reprinted from *Modern Medicine,* February and May 1895. The *most* thoroughly cooked starch found a lasting commercial product in Wonder Bread.

37. John H. Kellogg, *The Natural Diet of Man* (Battle Creek, Mich.: The Modern Medicine Publishing Co., 1923).

38. Carson, *Cornflake Crusade,* 125.

39. One of the most prominent articulations of technique for French cuisine is Jacques Pépin, *La Technique: The Fundamental Techniques of Cooking: An Illustrated Guide* (New York: Pocket Books, 1978).

40. Tim Cresswell, *Place: A Short Introduction* (Malden, Mass.: Blackwell, 2004).

41. Robert David Sack, *Homo Geographicus: A Framework for Action, Awareness, and Moral Concern* (Baltimore, Md.: Johns Hopkins University Press, 1997).

42. E. Kellogg, *Science in the Kitchen*, 72–73. *Multum in parvo* is Latin for "a great deal in a small space."

43. An exception to this in the context of the United States might include the phrase "Southern kitchen" to refer to a style of cooking and the implicitly suggested place of action. Among others examples, see the cookbook Sara Foster, *Sara Foster's Southern Kitchen* (New York: Random House, 2011). I hypothesize that the history of the linguistic separation of cuisine from kitchen is tied to changes in taste in the eighteenth century, wherein the vagaries of butchering and the application of heat (and other transformations of nature into art) were not suitable to the bourgeoisie. For a starting point, see Denise Gigante, *Taste: A Literary History* (New Haven: Yale University Press, 2005). This is a topic that needs much further investigation within cultural geography and food studies scholarship.

44. E. Kellogg, *Science in the Kitchen*, 63–64.

45. Note from John Kellogg to William Kellogg (typed manuscript, September 20, 1898), Michigan State University Archives and Historical Collections, John H. Kellogg Papers, East Lansing, Mich.

46. Stephen Mennell, *All Manners of Food: Eating and Taste in England and France from the Middle Ages to the Present* (New York: Blackwell, 1985).

47. See, by way of comparison, Arjun Appadurai, "How to Make a National Cuisine: Cookbooks in Contemporary India," *Comparative Studies in Society and History* 30, no. 1 (1988): 3–24.

48. James C. Jackson, *How to Treat the Sick without Medicine* (New York: Austin, Jackson and Co., 1871), 32 (emphasis in original).

49. Ibid., 34.

50. Nissenbaum, *Sex, Diet, and Debility in Jacksonian America*, 157.

51. John H. Kellogg, *Balanced Bills of Fare: Arranged with Reference to the Normal Daily Ration; And the Needs of Special Classes of Invalids* (Battle Creek, Mich.: Good Health Pub. Co., 1899).

52. See also Kendra Smith-Howard, *Pure and Modern Milk: An Environmental History since 1900* (New York: Oxford University Press, 2014).

53. Kelly Dowhower Karpa, *Bacteria for Breakfast: Probiotics for Good Health* (Victoria, B.C.: Trafford, 2003); and Natasha Trenev, *Probiotics: Nature's Internal Healers* (Garden City Park, N.Y.: Avery, 1998).

54. For a take on frozen concentrated orange juice, see Shane Hamilton, "Cold Capitalism: The Political Ecology of Frozen Concentrated Orange Juice," *Agricultural History* 77 (2003): 557–581.

4. EXTENDING THE DIGESTIVE SYSTEM INTO THE URBAN LANDSCAPE

1. Ben Anderson and John Wylie, "On Geography and Materiality," *Environment and Planning A* 41 (2009). See p. 321 for a discussion on the "assembling of materialities."

2. Arnold A. Clark, "The Sewerage and Drainage of Battle Creek," in *Report of Proceedings of the Sanitary Convention at Battle Creek: Supplement to the Annual Report of the State Board of Health,* Reprint No. 311 (1890): 37.

3. Jonathan Murdoch, "The Spaces of Actor-Network Theory," *Geoforum* 29, no. 4 (1998).

4. For example, see Scott Kirsch, "Cultural Geography III: Objects of Culture and Humanity, or, Re-'Thinging' the Anthropocene Landscape," *Progress in Human Geography* (2015); David Crouch, "Flirting with Space: Thinking Landscape Relationally," *Cultural Geographies* 17, no. 1 (2010).

5. For a thorough take on sewage treatment plants as ecosystems in themselves, as a hybrid between nature and industry, see Daniel Schneider, *Hybrid Nature: Sewage Treatment and the Contradictions of the Industrial Ecosystem* (Cambridge, Mass.: MIT Press, 2011).

6. Sandra Steingraber, *Living Downstream: An Ecologist Looks at Cancer and the Environment* (New York: Addison-Wesley Publishing Co., 1997); Nancy Langston, *Toxic Bodies: Hormone Disruptors and the Legacy of DES* (New Haven: Yale University Press, 2010); Gregg Mitman, *Breathing Space: How Allergies Shape Our Lives and Landscapes* (New Haven: Yale University Press, 2007).

7. Robert Koch, *Essays of Robert Koch,* trans. C. Carter (New York: Greenwood Press, 1987). "Lecture at the First Conference for Discussion of the Cholera Question" was first published in 1884. For this paragraph I reference pp. 151–153.

8. John Duffy, *The Sanitarians: A History of American Public Health* (Urbana: University of Illinois Press, 1990). See chapter 13, "Bacteriology Revolutionizes Public Health."

9. Battle Creek Department of Public Works, Records Division, Battle Creek, Mich. Sewer tap applications accessed by the author.

10. Order of the Common Council, *Charter of the City of Battle Creek, Michigan* (Battle Creek, Mich.: Gage and Sons, 1891), 28–29.

11. Personal communication with Charles "Ken" Kohs, utilities director, Battle Creek Department of Public Works, May 2009.

12. Larry B. Massie and Peter J. Schmitt, *Battle Creek: The Place behind the Products* (Woodland Hills, Calif.: Windsor Publications, 1984), 79.

13. Yet this development was slow. Even into the 1870s many large American cities only had ad hoc public health departments. See Martin V. Melosi, *The Sanitary City: Urban Infrastructure in America from Colonial Times to the Present* (Baltimore: Johns Hopkins University Press, 2000), 20.

14. Clark, "The Sewerage and Drainage of Battle Creek," 37.

15. *Good Health: A Journal of Hygiene* 22, no. 2 (February 1887).

16. Joel A. Tarr, *The Search for the Ultimate Sink: Urban Pollution in Historical Perspective* (Akron, Ohio: University of Akron Press, 1996), 115.

17. Melosi, *The Sanitary City*, 47.

18. Ella E. Kellogg, *Science in the Kitchen: The Principles of Healthful Cookery* (Battle Creek, Mich.: Modern Medicine Publishing Co., 1893), 61.

19. On the rich history of the emergence of bacteriology, see Gerald L. Geison, *The Private Science of Louis Pasteur* (Princeton, N.J.: Princeton University Press, 1995); Thomas D. Brock, *Robert Koch: A Life in Medicine and Bacteriology* (Washington, D.C.: American Society for Microbiology Press, 1999); and Nancy Tomes, *The Gospel of Germs: Men, Women, and the Microbe in American Life* (Cambridge, Mass.: Harvard University Press, 1998).

20. It's not a coincidence because Kellogg's program of health was in large part responsible for making the biggest industrial and population explosion that Battle Creek would ever see.

21. Erwin F. Smith, "Sewerage and Water-Supply: Discussion at a Sanitary Convention Held at Lansing, Mich.," reprinted from a supplement to the *Annual Report of the Michigan State Board of Health*, reprint no. 231 (1885): 105.

22. John H. Kellogg, "Sanitary Associations," *Seventh Annual Report of the Michigan State Board of Health* (1879): 92.

23. John H. Kellogg, "Decomposing Organic Matter," *Annual Report, Michigan State Board of Health*, reprint no. 128 (1882): 226.

24. Charles Bouchard, *Lectures on Auto-Intoxication in Disease; Or, Self-Poisoning of the Individual* (Philadelphia: F. A. Davis, 1894).

25. Kellogg was aware of Bouchard's work early on. For example, Kellogg translated and printed the following article, which cites and is based on Bouchard's work: A. Charrin and John H. Kellogg, trans., "The Poisons of the Urine," *Modern Medicine and Bacteriological Review* 3, no. 1 (1894). Kellogg also cites Bouchard in the first page of the preface of his Autointoxication; Or, Intestinal Toxemia" (Battle Creek, Mich.: Modern Medicine Pub. Co., 1922).

26. Alain Contrepois, "The Clinician, Germs and Infectious Diseases: The Example of Charles Bouchard in Paris," *Medical History* 46 (2002): 202.

27. John H. Kellogg, "Relation of Modern Physiological Chemistry to Vegetarianism," *Modern Medicine and Bacteriological Review* 3, no. 3 (1894): 58–59.

28. John H. Kellogg, *Constipation: How to Fight It* (Battle Creek, Mich.: Good Health Publishing Co., 1913), 3.

29. *Diet Helps for Doctors* (Battle Creek, Mich.: Diet Service Department of the Battle Creek Food Co., 1928); quotes in this paragraph pp. 32–34.

30. Earl Henry Rathbun, "Millions in Every Bushel" (typed memoir, University of Michigan Bentley Historical Library, ca. 1920).

31. For a more general discussion on how body parts became seen as independent entities in modernity, see Ian Hacking, "Our Neo-Cartesian Bodies in Parts," *Critical Inquiry* 34 (2007): 78–105.

32. John H. Kellogg, *Rational Hydrotherapy: A Manual of the Physiological and Therapeutic Effects of Hydriatic Procedures, and the Technique of Their Application in the Treatment of Disease* (Philadelphia: F. A. Davis Co., 1901): 891.

33. Order of the Common Council, *Charter,* 30–31.

34. *Good Health: A Journal of Hygiene* 22, no. 1 (January 1887): 31.

35. Hugh E. McKelvey, *Some History of Beautiful Goguac Lake* (ca. 1995) (manuscript held at the Willard Library, Helen Warner Branch, Battle Creek, Mich.): 1.

36. John H. Kellogg, *The Itinerary of a Breakfast: A Popular Account of the Travels of a Breakfast through the Food Tube and of the Ten Gates and Several Stations through Which It passes, Also of the Obstacles Which It Sometimes Meets* (New York: Funk and Wagnalls Company, 1918): 18–22.

37. See, by way of comparison, Simon Marvin and Will Medd, "Metabolisms of Obe*city:* Flows of Fat through Bodies, Cities, and Sewers," *Environment and Planning A* 38 (2006).

5. THE SYSTEMATIZATION OF AGRICULTURE

1. John H. Kellogg, *The Stomach: Its Disorders and How to Cure Them* (Battle Creek, Mich.: Modern Medicine Publishing Co., 1896), 94.

2. The source for much of the sanitarium's wheat was local, though the geographical extent of "local" in this case in unknown. Gerald Carson wrote, for example, "W. K. Kellogg . . . weighed in the local wheat at the Sanitas wagon scales." See Gerald Carson, *Cornflake Crusade* (New York: Rinehart and Co., 1957), 197.

3. Kellogg, *The Stomach,* 19–20.

4. For good examples of how Kellogg applied the language of quantification to the digestive system, see John H. Kellogg, "An Exact Method for Determining the Capacity of the Stomach and the Amount of Residual Contents," *Modern Medicine and Bacteriological Review* 5, no. 7 (1896); John H. Kellogg, "Cancer of the Stomach" (paper read before the Michigan State Medical Society, 1897).

5. Erik Swyngedouw, "Circulations and Metabolisms: (Hybrid) Natures and (Cyborg) Cities," in *Technonatures: Environments, Technologies, Spaces and Places in the Twenty-First Century,* ed. C. Wilbert and D. F. White (Waterloo: Wilfrid Laurier University Press, 2009), 65–66.

6. S. J. Youngman, "What Crop Will Pay?" *Michigan Farmer and State Journal of Agriculture* 23, no. 18 (1892).

7. Alan I. Marcus, *Agricultural Science and the Quest for Legitimacy: Farmers, Agricultural Colleges, and Experiment Stations, 1870–1890* (Ames: Iowa State University Press, 1985), 5.

8. Quote is from ibid., 14. See also Deborah Fitzgerald, *Every Farm a Factory: The Industrial Ideal in American Agriculture* (New Haven: Yale University Press, 2003).

9. Ibid., 109.

10. Kellogg, *The Stomach,* 18.

11. Ibid., 226.

12. John H. Kellogg, "The Treatment of Hyperpepsia," *Modern Medicine and Bacteriological Review* 3, no. 4 (1894): 94. The word *vitiate* generally means "to reduce the quality of," though it can also mean "to corrupt morally." This twin denotation is congruent with Kellogg's nation-building-through-stomach-building agenda.

13. Naomi Aronson, "Nutrition as a Social Problem: A Case Study of Entrepreneurial Strategy in Science," *Social Problems* 29, no. 5 (1982): 474–487.

14. James C. Scott, *Seeing Like a State: How Certain Schemes to Improve the Human Condition Have Failed* (New Haven: Yale University Press, 1998), 262–263.

15. Margaret W. Rossiter, *The Emergence of Agricultural Science: Justus Liebig and the Americans, 1840–1880* (New Haven: Yale University Press, 1975).

16. Marcus, *Agricultural Science and the Quest for Legitimacy.*

17. Dana G. Dalrymple, "The Smithsonian Bequest, Congress, and Nineteenth-Century Efforts to Increase and Diffuse Agricultural Knowledge in the United States," *Agricultural History Review* 57, no. 2 (2009): 209.

18. Humphry Davy, *Elements of Agricultural Chemistry,* 2nd ed. (London: Longman, Hurst, Rees, Orme and Brown, 1814).

19. W. A. Shenstone, *Justus Von Liebig: His Life and Work (1803–1873)* (New York: MacMillan, 1895), 88.

20. Justus von Liebig, *Organic Chemistry in Its Applications to Agriculture and Physiology* (London: Taylor and Walton, 1840).

21. Rossiter, *The Emergence of Agricultural Science,* 25. Even in 2016 botanists continue to discover how plants circulate minerals and energy in relation to the soil. See Monika A. Gorzelak, Amanda K. Asay, Brian J. Pickles, and Suzanne W. Simard, "Inter-Plant Communication through Mycorrhizal Networks Mediates

Complex Adaptive Behaviour in Plant Communities," *AoB Plants* 7 (2015), doi: 10.1093/aobpla/plv050.

22. Dalrymple, "The Smithsonian Bequest," 213. Albrecht Thaer's agricultural school, founded in 1806 near Berlin, is one mark of the beginning of modern agricultural science, serving as the model for agricultural education until the 1840s. On twelve hundred acres students learned about farming from both theoretical and practical perspectives. See also Karl Kautsky, *The Agrarian Question*, vol. 1., trans. P. Burgess, (Winchester, Mass.: Zwan Publications, 1988; first published in 1899), 56–57.

23. Rossiter, *The Emergence of Agricultural Science*, 25.

24. Kautsky, *The Agrarian Question*, 52–53.

25. Alfred C. True, *A History of Agricultural Experimentation and Research in the United States, 1607–1925* (Washington, D.C.: U.S. Department of Agriculture Miscellaneous Publication no. 251, Government Printing Office, 1937), 16.

26. Dalrymple, "The Smithsonian Bequest," 230. See also Paul E. Waggoner, "Research and Education in American Agriculture," *Agricultural History* 50, no. 1 (1976): 233.

27. Charles E. Rosenberg, "Science, Technology, and Economic Growth: The Case of the Agricultural Experiment Station Scientist, 1875–1914," in *Nineteenth-Century American Science: A Reappraisal*, ed. G. H. Daniels (Evanston, Ill.: Northwestern University Press, 1972), 182.

28. Marcus, *Agricultural Science and the Quest for Legitimacy*, 23.

29. Ibid., 23–24.

30. True, *A History of Agricultural Experimentation*, 137.

31. Liebig, *Organic Chemistry*, vii.

32. Dorothy Schwieder, "Agricultural Issues in the Middle West, 1865–1910," in *Agriculture and National Development: Views on the Nineteenth Century*, ed. L. Ferleger (Ames: Iowa State University Press, 1990), 107–108.

33. Marcus, *Agricultural Science and the Quest for Legitimacy*, ix.

34. True, *A History of Agricultural Experimentation*, 130.

35. Ibid., 175.

36. Rossiter, *The Emergence of Agricultural Science*, 127–128.

37. Samuel W. Johnson, *How Crops Grow: A Treatise on the Chemical Composition, Structure and Life of the Plant, for Students of Agriculture* (New York: Orange Judd Co., 1900), 2.

38. Aronson, "Nutrition as a Social Problem," 475.

39. Ibid., 476.

40. Rossiter, *The Emergence of Agricultural Science*, 163.

41. Justus von Liebig, *Researches on the Chemistry of Food, and the Motion of the Juices in the Animal Body* (Lowell, MA: Daniel Bixby, 1848).

42. Paul E. Howe, "Liebig and the Chemistry of Animal Nutrition," in *Liebig and After Liebig: A Century of Progress in Agricultural Chemistry*, ed. F. R.

Moulton (Washington, D.C.: American Association for the Advancement of Science, 1942), 40.

43. Wilbur O. Atwater, *Methods and Results of Investigations on the Chemistry and Economy of Food: Bulletin No. 21* (Washington, D.C.: Government Printing Office; U.S. Department of Agriculture Office of Experiment Stations, 1895); Wilbur O. Atwater and Chas D. Woods, *The Chemical Composition of American Food Materials: Bulletin No. 28* (Washington, D.C.: Government Printing Office; U.S. Department of Agriculture Office of Experiment Stations, 1896); Wilbur O. Atwater and F. G. Benedict, *Experiments on the Metabolism of Matter and Energy in the Human Body* (Washington, D.C.: Government Printing Office, 1902).

44. Norman W. Storer, *Science and Scientists in an Agricultural Research Organization: A Sociological Study* (New York: Arno Press, 1980).

45. Kenneth Carpenter, "The Life and Times of W. O. Atwater (1844–1907)," *Journal of Nutrition* 124 (1994).

46. Michel Foucault, *The History of Sexuality, Volume I: An Introduction*, trans. R. Hurley (New York: Vintage Books, 1990), 140–141; Carole A. Davis, Patricia Britten, and Esther Myers, "Past, Present, and Future of the Food Guide Pyramid," *Journal of the American Dietetic Association* 101, no. 8 (2001).

47. John H. Kellogg, *The Battle Creek Sanitarium System: History, Organization, Methods* (Battle Creek, Mich.: Gage Printing Co., 1908), 117–119.

48. Atwater and Woods, *The Chemical Composition of American Food Materials*, 8.

49. F. Riddell, "The Other Side of the Food Question," *Michigan Farmer and State Journal of Agriculture* 23, no. 29 (1892).

50. Edward W. Bemis, "The Discontent of the Farmer," *The Journal of Political Economy* 1, no. 2 (1893).

51. Richard Walker, *The Conquest of Bread: 150 Years of Agribusiness in California* (New York: New Press, 2004), 6–7.

52. Edwin Willits, *Industrial Education*, in Department of Agriculture, *Miscellaneous Special Report No. 9: Proceedings of a Convention of Delegates from Agricultural Colleges and Experiment Stations Held at the Department of Agriculture* (Washington, D.C.: Government Printing Office, 1885), 62.

53. David Goodman, Bernardo Sorj, and John Wilkinson, *From Farming to Biotechnology: A Theory of Agro-Industrial Development* (New York: Blackwell, 1987), 1.

54. Donald L. Winters, "The Economics of Midwestern Agriculture, 1865–1900," in *Agriculture and National Development: Views on the Nineteenth Century*, ed. L. Ferleger (Ames: Iowa State University Press, 1990), 81.

55. Ibid.

56. Frederick J. Turner, "The Significance of the Frontier in American History," in *The Frontier in American History* (Malabar, Fla.: R. E. Krieger, 1975).

57. Thorstein Veblen, "The Food Supply and the Price of Wheat," *The Journal of Political Economy* 1, no. 3 (1893): 379.

58. C. F. Emerick, "An Analysis of Agricultural Discontent in the United States. I.," *Political Science Quarterly* 11, no. 3 (1896): 436.

59. Willard Cochrane, *The Development of American Agriculture: A Historical Analysis* (Minneapolis: University of Minnesota Press, 1979), 195–196.

60. Ibid., 196.

61. Kautsky, *The Agrarian Question,* 44.

62. Ibid., 46.

63. Bernice B. Lowe, *Tales of Battle Creek* (Battle Creek, Mich.: Albert L. and Louise B. Miller Foundation, 1976), 64.

64. Harold A. Hooker, "The History and Development of the Oliver Corporation in Battle Creek, Michigan" (1952), Michigan State University Archives and Historical Collections, East Lansing, Mich. The Oliver Corporation is the name of what used to be known as the Nichols & Shepard Corporation.

65. Ibid., 4.

66. Edith Butler, "Growing Up on a Michigan Farm in the 1890s," *Michigan History* 68, no. 2 (1984), 14.

67. *Nichols and Shepard Co. Vibrator Threshing Machine catalog* (ca. 1890), Michigan Trade Catalogs collection, Box 1: Agricultural machinery, Bentley Historical Library, University of Michigan.

68. Ibid.

69. Kautsky, *The Agrarian Question,* 21.

70. Gladys L. Baker, V. Rasmussen, and J. M. Porter, *Century of Service: The First One Hundred Years of the Department of Agriculture* (Washington, D.C.: United States Government Printing Office, 1963). From Jack R. Kloppenburg, *First the Seed: The Political Economy of Plant Biotechnology, 1492–2000* (New York: Cambridge University Press, 1988), 59.

71. Alfred W. Crosby, *Ecological Imperialism: The Biological Expansion of Europe, 900–1900* (New York: Cambridge University Press, 1986).

72. Kloppenburg, *First the Seed,* 60.

73. "Seed Wheat," *Michigan Farmer and State Journal of Agriculture* 23, no. 35 (1892).

74. Ibid.

75. This quote and much of the information in this paragraph are from Kloppenburg, *First the Seed,* 62–65.

76. Alan L. Olmstead and Paul W. Rhode, *Creating Abundance: Biological Innovation and American Agricultural Development* (New York: Cambridge University Press, 2008), 28.

77. Patsy Gerstner, "The Temple of Health: A Pictorial History of the Battle Creek Sanitarium," *Caduceus* 12, no. 2 (1996): 33–34.

78. Note from W.K. Kellogg to A.S. Kellogg (May 5, 1899), John H. Kellogg Papers, "Food Experiments" folders, Michigan State University Archives and Historical Collections, East Lansing, Mich.

79. Richard W. Schwarz, "John Harvey Kellogg: American Health Reformer" (dissertation, History, University of Michigan, Ann Arbor, 1965), p. 184.

80. Kellogg, *The Battle Creek Sanitarium System*, 138 (emphasis added).

81. Bruno Latour, *Pandora's Hope: Essays on the Reality of Science Studies* (Cambridge, Mass.: Harvard University Press, 1999), 24–31.

6. BREAKFAST CEREAL IN THE TWENTIETH CENTURY

1. John H. Kellogg, *The Battle Creek Sanitarium System: History, Organization, Methods* (Battle Creek, Mich.: Gage Printing Co., 1908), 137.

2. Gerald Carson, *Cornflake Crusade* (New York: Rinehart and Co., 1957), 108.

3. *Michigan State Gazetteer and Business Directory*, vol. XXIV (Detroit: R.L. Polk and Co., 1905–1906); *Battle Creek, Albion, Marshall, Etc. Directory* (Detroit: R.L. Polk and Co., 1896).

4. Larry B. Massie and Peter J. Schmitt, *Battle Creek: The Place behind the Products* (Woodland Hills, Calif.: Windsor Publications, 1984), 7.

5. Battle Creek Sesquicentennial Book Committee, *Battle Creek 150: Sesquicentennial, 1831–1981* (Battle Creek, Mich.: Embossing Printers, 1980), 25.

6. "101 and We've Just Begun: Cereal Manufacturing Companies in Battle Creek," *Heritage Battle Creek: A Journal of Local History* 2 (Spring 1992).

7. Horace B. Powell, *The Original Has This Signature—W.K. Kellogg: The Story of a Pioneer in Industry and Philanthropy* (Battle Creek, Mich.: W.K. Kellogg Foundation, 1989; originally published in 1956), 92–93.

8. Battle Creek Sesquicentennial Book Committee, *Battle Creek 150*, 8.

9. Carson, *Cornflake Crusade*, 197.

10. Ibid., 198.

11. Ibid., 199.

12. Ibid., 207.

13. Willard Cochrane, *The Development of American Agriculture: A Historical Analysis* (Minneapolis: University of Minnesota Press, 1979), 91–92.

14. Edward W. Bemis, "The Discontent of the Farmer," *The Journal of Political Economy* 1, no. 2 (1893).

15. William Cronon, *Nature's Metropolis: Chicago and the Great West* (New York: W.W. Norton, 1991), 145.

16. Sidney W. Mintz, *Sweetness and Power: The Place of Sugar in Modern History* (New York: Penguin Books, 1986).

17. Ibid., 196.

18. The list of advertising strategies is from Carson, *Cornflake Crusade*, 209–210.

19. Television advertisement viewed by the author on May 14, 2008, at the Paley Center for Media in Beverly Hills, Calif.

20. Upton Sinclair, *The Jungle* (New York: Penguin Books, 1906).

21. Andrew F. Smith, *Eating History: 30 Turning Points in the Making of American Cuisine* (New York: Columbia University Press, 2009).

22. Powell, *The Original Has This Signature*, 182.

23. Ian Hutchby, "Technologies, Texts and Affordances," *Sociology* 35, no. 2 (2001).

24. Powell, *The Original Has This Signature*, 184.

25. E. Melanie DuPuis, *Nature's Perfect Food: How Milk Became America's Drink* (New York: New York University Press, 2002).

26. William Prout, *Chemistry, Meteorology, and the Function of Digestion: Considered with Reference to Natural Theology* (London: W. Pickering, 1834). From DuPuis, *Nature's Perfect Food*, 69.

27. DuPuis, *Nature's Perfect Food*, 74.

28. John H. Kellogg, *The Household Monitor of Health* (Battle Creek, Mich.: Good Health Publishing Co., 1891), 126.

29. John H. Kellogg, *The Stomach: Its Disorders and How to Cure Them* (Battle Creek, Mich.: Modern Medicine Publishing Co., 1896), 35.

30. T. J. Jackson Lears, *No Place of Grace: Antimodernism and the Transformation of American Culture, 1880–1920* (New York: Pantheon, 1981), xiii.

31. David Lowenthal, *The Past is a Foreign Country* (New York: Cambridge University Press, 1985), 7.

32. James Whorton, *Inner Hygiene: Constipation and the Pursuit of Health in Modern Society* (New York: Oxford University Press, 2000), 31.

33. Donald Worster, *Nature's Economy: A History of Ecological Ideas* (New York: Cambridge University Press, 1977).

34. H. Thompson Straw, "Battle Creek: A Study in Urban Geography" (unpublished thesis, University of Michigan, Ann Arbor, 1938), 83.

35. Whorton, *Inner Hygiene*, 32 (emphasis added).

36. L. J. Rather, "The 'Six Things Non-Natural': A Note on the Origins and Fate of a Doctrine and a Phrase," *Clio Medica* 3 (1968): 337.

37. Michel Serres, *The Parasite* (Baltimore: Johns Hopkins University Press, 1982).

38. David Nye, ed. *Technologies of Landscape: From Reaping to Recycling* (Amherst, Mass.: University of Massachusetts Press, 1999).

39. For a related line of thinking, see Jennifer Gabrys, "Sink: The Dirt of Systems," *Environment and Planning D: Society and Space* 27 (2009).

40. Ibid.

41. Bruno Latour, *Reassembling the Social: An Introduction to Actor-Network-Theory* (New York: Oxford University Press, 2005), 173.

42. Rogers Brubaker, *Nationalism Reframed: Nationhood and the National Question in the New Europe* (New York: Cambridge University Press, 1996).

43. E. Melanie DuPuis, "A Place Called Enteria: The Gastro-Geopolitics of the Colon" (paper read at the Association of American Geographers annual meeting, March 22–27, 2009, Las Vegas, Nev.).

44. Denis Cosgrove, "Prospect, Perspective and the Evolution of the Landscape Idea," *Transactions of the Institute of British Geographers* 10, no. 1 (1985).

45. Nicholas Entrikin makes a similar observation about the concept of "place" in J. Nicholas Entrikin, *The Betweenness of Place: Towards a Geography of Modernity* (Baltimore: Johns Hopkins University Press, 1991).

46. See also Michel de Certeau, *The Practice of Everyday Life* (Berkeley: University of California Press, 1984).

47. A thorough introduction to the problems this duality and others have generated in human geography is Paul J. Cloke and Ron Johnston, *Spaces of Geographical Thought: Deconstructing Human Geography's Binaries* (Thousand Oaks, Calif.: Sage, 2005).

48. For geographer John Wylie's accounts of "being in" and "touching" landscape, see John Wylie, "Landscape and Phenomenology," in *The Routledge Companion to Landscape Studies,* ed. P. Howard, I. Thompson, and E. Waterton (New York: Routledge, 2013); and John Wylie, "A Single Day's Walking: Narrating Self and Landscape on the South West Coast Path," *Transactions of the Institute of British Geographers* 30 (2005).

49. Two works by Emily Martin are excellent for thinking about bodies as a process, or constant becoming: Emily Martin, "Fluid Bodies, Managed Nature," in *Remaking Reality: Nature at the millennium,* ed. B. Braun and N. Castree (New York: Routledge, 1998); and Emily Martin *Flexible Bodies: Tracking Immunity in American Culture from the Days of Polio to the Age of AIDS* (Boston: Beacon Press, 1994).

50. "The material interface between the body and the city is perhaps most strikingly manifested in the physical infrastructure that links the human body to vast technological networks. . . . The modern home, for example, has become a complex exoskeleton for the human body with a provision of water, warmth, light and other essential needs." From Matthew Gandy, "Cyborg Urbanization: Complexity and Monstrosity in the Contemporary City," *International Journal of Urban and Regional Research* 29, no. 1 (2005): 28.

Bibliography

"About Battle Creek." *Headlight: A Periodical Devoted to the Interests of Railroads and Railroad Centers,* May 1895, 5–46.

Alaimo, Stacy. "Thinking as the Stuff of the World." *O-Zone: A Journal of Object-Oriented Studies* 1 (2014): 13–21.

Anderson, Ben, and John Wylie. "On Geography and Materiality." *Environment and Planning A* 41 (2009): 318–335.

Anderson, Heather Arndt. *Breakfast: A History.* Lanham, Md.: Rowman and Littlefield, 2013.

Anderson, Wes (director). *The Grand Budapest Hotel.* United States: Fox Searchlight, 2014.

Appadurai, Arjun. "How to Make a National Cuisine: Cookbooks in Contemporary India." *Comparative Studies in Society and History* 30, no. 1 (1988): 3–24.

Aronson, Naomi. "Nutrition as a Social Problem: A Case Study of Entrepreneurial Strategy in Science." *Social Problems* 29, no. 5 (1982): 474–487.

Atwater, W. O. *Methods and Results of Investigations on the Chemistry and Economy of Food: Bulletin no. 21.* Washington, D.C.: Government Printing Office; U.S. Department of Agriculture Office of Experiment Stations, 1895.

Atwater, W. O., and F. G. Benedict. *Experiments on the Metabolism of Matter and Energy in the Human Body.* Washington, D.C.: Government Printing Office, 1902.

Atwater, W. O., and Chas D. Woods. *The Chemical Composition of American Food Materials: Bulletin no. 28*. Washington, D.C.: Government Printing Office; U.S. Department of Agriculture Office of Experiment Stations, 1896.

Baker, Gladys L., V. Rasmussen, and J. M. Porter. *Century of Service: The First One Hundred Years of the Department of Agriculture*. Washington, D.C.: United States Government Printing Office, 1963.

Baker, S. M. "Nursing in Homes, Private Hospitals, and Sanitariums." *Modern Medicine and Bacteriological Review* 3, no. 4 (1894): 77–82.

Battle Creek, Albion, Marshall, Etc. Directory. Detroit: R. L. Polk and Co., 1896.

Battle Creek Sesquicentennial Book Committee. *Battle Creek 150: Sesquicentennial, 1831–1981*. Battle Creek, Mich.: Embossing Printers, 1980.

Beardsley, Edward H. *The Rise of the American Chemistry Profession, 1850–1900*. Gainesville: University of Florida Press, 1964.

Beaumont, William. *Experiments and Observations on the Gastric Juice, and the Physiology of Digestion*. Plattsburgh, N.Y.: F. P. Allen, 1833.

Beebe, Alan R. "History of the Battle Creek Sanitarium, 1866–1903." Unpublished Thesis, Kalamazoo College, Kalamazoo, Mich., 1949.

Belasco, Warren J. *Appetite for Change: How the Counterculture Took on the Food Industry*. 2nd ed. Ithaca, N.Y.: Cornell University Press, 2007.

Bemis, Edward W. "The Discontent of the Farmer." *The Journal of Political Economy* 1, no. 2 (1893): 193–213.

Biltekoff, Charlotte. *Eating Right in America: The Cultural Politics of Food and Health*. Durham: Duke University Press, 2013.

Bouchard, Charles. *Lectures on Auto-Intoxication in Disease; Or, Self-Poisoning of the Individual*. Philadelphia: F. A. Davis, 1894.

Bowker, Geoffrey, and Susan Leigh Star. *Sorting Things Out: Classification and Its Consequences*. Cambridge, Mass.: MIT Press, 1999.

Boyle, T. Coraghessan. *The Road to Wellville*. New York: Penguin, 1993.

Brock, Thomas D. *Robert Koch: A Life in Medicine and Bacteriology*. Washington, D.C.: American Society for Microbiology Press, 1999.

Brubaker, Rogers. *Nationalism Reframed: Nationhood and the National Question in the New Europe*. New York: Cambridge University Press, 1996.

Bruce, Scott, and Bill Crawford. *Cerealizing America: The Unsweetened Story of American Breakfast Cereal*. Boston: Faber and Faber, 1995.

Buck, Daniel, Christina Getz, and Julie Guthman. "From Farm to Table: The Organic Vegetable Commodity Chain of Northern California." *Sociologia Ruralis* 37, no. 1 (1997): 3–20.

Bull, Malcolm, and Keith Lockhart. *Seeking Sanctuary: Seventh-day Adventism and the American Dream*. Bloomington: Indiana University Press, 2007.

Butler, Edith. "Growing Up on a Michigan Farm in the 1890s." *Michigan History* 68, no. 2 (1984).

Caldwell, Melissa L. "Digestive Politics in Russia: Feeling the Sensorium beyond the Palate." *Food and Foodways* 22, nos. 1–2 (2014): 112–135.

Calhoun County Souvenir: Commemorating the Battle Creek Journal's Fiftieth Anniversary: Calhoun County, Her Industries and Sketches of Her Representative Citizens. Battle Creek, Mich.: Battle Creek Journal, 1901.

Carpenter, Kenneth. "The Life and Times of W. O. Atwater (1844–1907)." *Journal of Nutrition* 124 (1994): 1707S–1714S.

Carson, Gerald. *Cornflake Crusade.* New York: Rinehart and Co., 1957.

Castree, Noel, and Catherine Nash. "Editorial: Posthuman Geographies." *Social and Cultural Geography* 7, no. 4 (2006): 501–504.

Cayleff, Susan E. *Wash and Be Healed: The Water-Cure Movement and Women's Health.* Philadelphia: Temple University Press, 1987.

Charrin, A., and John H. Kellogg. "The Poisons of the Urine." *Modern Medicine and Bacteriological Review* 3, no. 1 (1894): 10–12.

Claridge, R. T. *Hydropathy; Or, the Cold Water Cure, as Practiced by Vincent Priessnitz, at Graefenberg, Silesia, Austria.* 8th ed. London: James Madden and Co., 1843.

Clark, A. Arnold. "The Sewerage and Drainage of Battle Creek." In *Report of Proceedings of the Sanitary Convention at Battle Creek: Supplement to the Annual Report of the State Board of Health,* Reprint No. 311 (1890).

Cloke, Paul J., and Ron Johnston. *Spaces of Geographical Thought: Deconstructing Human Geography's Binaries.* Thousand Oaks, Calif.: Sage, 2005.

Cochrane, Willard. *The Development of American Agriculture: A Historical Analysis.* Minneapolis: University of Minnesota Press, 1979.

Contrepois, Alain. "The Clinician, Germs and Infectious Diseases: The Example of Charles Bouchard in Paris." *Medical History* 46 (2002): 197–220.

Coole, Diana, and Samantha Frost. "Introducing the New Materialisms." In *New Materialisms: Ontology, Agency, and Politics,* edited by D. Coole and S. Frost. Durham, N.C.: Duke University Press, 2010.

Corbin, Alain. *The Foul and the Fragrant: Odor and the French Social Imagination.* Cambridge, Mass.: Harvard University Press, 1986.

Cosgrove, Denis. "Prospect, Perspective and the Evolution of the Landscape Idea." *Transactions of the Institute of British Geographers* 10, no. 1 (1985): 45–62.

Cresswell, Tim. *Place: A Short Introduction.* Malden, Mass.: Blackwell, 2004.

Cronon, William. *Nature's Metropolis: Chicago and the Great West.* New York: W. W. Norton, 1991.

Crosby, Alfred W. *Ecological Imperialism: The Biological Expansion of Europe, 900–1900.* New York: Cambridge University Press, 1986.

Crouch, David. "Flirting with Space: Thinking Landscape Relationally." *Cultural Geographies* 17, no. 1 (2010): 5–18.

Dalrymple, Dana G. "The Smithsonian Bequest, Congress, and Nineteenth-Century Efforts to Increase and Diffuse Agricultural Knowledge in the United States." *Agricultural History Review* 57, no. 2 (2009): 207–235.

Davis, Carole A., Patricia Britten, and Esther Myers. "Past, Present, and Future of the Food Guide Pyramid." *Journal of the American Dietetic Association* 101, no. 8 (2001): 881–885.

Davy, Humphry. *Elements of Agricultural Chemistry*. 2nd ed. London: Longman, Hurst, Rees, Orme and Brown, 1814.

de Certeau, Michel. *The Practice of Everyday Life*. Berkeley: University of California Press, 1984.

de la Peña, Carolyn. "Recharging at the Fordyce: Confronting the Machine and Nature in the Modern Bath." *Technology and Culture* 40, no. 4 (1999): 746–769.

Deutsch, Ronald M. *The Nuts among the Berries: An Exposé of America's Food Fads*. New York: Ballantine Books, 1967.

Diamond, Harvey, and Marilyn Diamond. *Fit for Life*. New York: Warner Books, 1985.

"Diet Helps for Doctors." Battle Creek, Mich.: Diet Service Department of the Battle Creek Food Co., 1928.

"Dr. Kellogg Recalls Early Sanitarium Diet Experiments." *Battle Creek Moon-Journal* 24 (November 21, 1938).

Duffy, John. *The Sanitarians: A History of American Public Health*. Urbana: University of Illinois Press, 1990.

DuPuis, E. Melanie. *Nature's Perfect Food: How Milk Became America's Drink*. New York: New York University Press, 2002.

———. *Dangerous Digestion: The Politics of American Dietary Advice*. Oakland: University of California Press, 2015.

———. "A Place Called Enteria: The Gastro-Geopolitics of the Colon." Paper read at the Association of American Geographers annual meeting, March 22–27, 2009, Las Vegas, NV.

Emerick, C. F. "An Analysis of Agricultural Discontent in the United States. I." *Political Science Quarterly* 11, no. 3 (1896): 433–463.

Entrikin, J. Nicholas. *The Betweenness of Place: Towards a Geography of Modernity*. Baltimore, Md.: Johns Hopkins University Press, 1991.

Fast, Robert B., and Elwood F. Caldwell. *Breakfast Cereals and How They Are Made*. 2nd ed. St. Paul, Minn.: American Association of Cereal Chemists, 2000.

Fitzgerald, Deborah. *Every Farm a Factory: The Industrial Ideal in American Agriculture*. New Haven, Conn.: Yale University Press, 2003.

Foley, Ronan. *Healing Waters: Therapeutic Landscapes in Historic and Contemporary Ireland*. Burlington, Vt.: Ashgate, 2010.

———. "Performing Health in Place: The Holy Well as a Therapeutic Assemblage." *Health and Place* 17, no. 2 (2011): 470–479.

Foster, Sara. *Sara Foster's Southern Kitchen*. New York: Random House, 2011.

Foucault, Michel. *The History of Sexuality, Volume I: An Introduction*. Translated by R. Hurley. New York: Vintage Books, 1990. Originally published in 1976.

———. *Discipline and Punish: The Birth of the Prison*. Translated by A. Sheridan. New York: Vintage Books, 1995. Originally published in 1975.

Gabrys, Jennifer. "Sink: The Dirt of Systems." *Environment and Planning D: Society and Space* 27 (2009): 666–681.

Gandy, Matthew. "Cyborg Urbanization: Complexity and Monstrosity in the Contemporary City." *International Journal of Urban and Regional Research* 29, no. 1 (2005): 26–49.

Geison, Gerald L. *The Private Science of Louis Pasteur*. Princeton, N.J.: Princeton University Press, 1995.

Gerstner, Patsy. "The Temple of Health: A Pictorial History of the Battle Creek Sanitarium." *Caduceus* 12, no. 2 (1996): 3–99.

Gesler, Wilbert M. *Healing Places*. New York: Rowman and Littlefield, 2003.

Gigante, Denise. *Taste: A Literary History*. New Haven, Conn.: Yale University Press, 2005.

Goodman, David, Bernardo Sorj, and John Wilkinson. *From Farming to Biotechnology: A Theory of Agro-Industrial Development*. New York: Blackwell, 1987.

Graham, Roy E. *Ellen G. White: Co-Founder of the Seventh-day Adventist Church*. New York: Peter Lang, 1985.

Graham, Sylvester. *A Treatise on Bread, and Bread-Making*. Boston: Light and Stearns, 1837.

Gregory, Derek. "Paradigm." In *The Dictionary of Human Geography*, edited by R.J. Johnston, D. Gregory, G. Pratt, and M. Watts. Malden, Mass.: Blackwell, 2000.

Guthman, Julie. *Weighing In: Obesity, Food Justice, and the Limits of Capitalism*. Berkeley: University of California Press, 2011.

Hacking, Ian. "Our Neo-Cartesian Bodies in Parts." *Critical Inquiry* 34 (2007): 78–105.

Hamilton, Shane. "Cold Capitalism: The Political Ecology of Frozen Concentrated Orange Juice." *Agricultural History* 77 (2003): 557–581.

Harman, Graham. *Guerrilla Metaphysics: Phenomenology and the Carpentry of Things*. Chicago: Open Court, 2005.

Hartshorne, Richard. *The Nature of Geography: A Critical Survey of Current Thought in the Light of the Past*. Lancaster, Pa.: The Association of American Geographers, 1939.

Hayles, N. Katherine. *How We Became Posthuman: Virtual Bodies in Cybernetics, Literature, and Informatics.* Chicago: University of Chicago Press, 1999.

"A Health Resort." *The Sunday Inter Ocean* 23, no. 134 (1894): 21.

Hooker, Harold A. "The History and Development of the Oliver Corporation in Battle Creek, Michigan" (1952). Michigan State University Archives and Historical Collections, East Lansing, Mich.

Howe, Paul E. "Liebig and the Chemistry of Animal Nutrition." In *Liebig and After Liebig: A Century of Progress in Agricultural Chemistry,* edited by F. R. Moulton. Washington, D.C.: American Association for the Advancement of Science, 1942.

Hudson, Robert P. "Theory and Therapy: Ptosis, Stasis, and Autointoxication." *Bulletin of the History of Medicine* 63, no. 3 (1989): 392–413.

Hutchby, Ian. "Technologies, Texts and Affordances." *Sociology* 35, no. 2 (2001): 441–456.

Jackson, James C. *How to Treat the Sick without Medicine.* New York: Austin, Jackson and Co., 1871.

Jackson, La Scienya, Stanley Dudrick, and Bauer Sumpio. "John Harvey Kellogg: Surgeon, Inventor, Nutritionist (1852–1943)." *Journal of the American College of Surgeons* 199, no. 5 (2004): 817–821.

Johnson, Samuel W. *How Crops Grow: A Treatise on the Chemical Composition, Structure and Life of the Plant, for Students of Agriculture.* New York: Orange Judd Co., 1900.

Jons, Heike. "Dynamic Hybrids and the Geographies of Technoscience: Discussing Conceptual Resources beyond the Human/Non-Human Binary." *Social and Cultural Geography* 7, no. 4 (2006): 559–580.

Karpa, Kelly Dowhower. *Bacteria for Breakfast: Probiotics for Good Health.* Victoria, B.C.: Trafford, 2003.

Kautsky, Karl. *The Agrarian Question,* vol 1. Translated by P. Burgess. Winchester, Mass.: Zwan Publications, 1988. First published in 1899.

Kellogg, Ella E. *Science in the Kitchen: The Principles of Healthful Cookery.* Battle Creek, Mich.: Modern Medicine Publishing Co., 1893.

Kellogg, John H. *Autointoxication; Or, Intestinal Toxemia.* Battle Creek, Mich.: Modern Medicine Pub. Co., 1922.

———. *Balanced Bills of Fare: Arranged with Reference to the Normal Daily Ration; And the Needs of Special Classes of Invalids.* Battle Creek, Mich.: Good Health Pub. Co., 1899.

———. *The Battle Creek Sanitarium System: History, Organization, Methods.* Battle Creek, Mich.: Gage Printing Co., 1908.

———. "Cancer of the Stomach." Paper read before the Michigan State Medical Society, 1897, 1–11.

———. *Constipation: How to Fight It.* Battle Creek, Mich.: Good Health Publishing Co., 1913.

———. "Decomposing Organic Matter." Paper read before the Sanitary Convention, Michigan State Board of Health, 1882.

———. "Dependable Methods of Producing and Maintaining an Aciduric Intestinal Flora." Reprinted from *the Medical Journal and Record* (1929).

———. "An Exact Method for Determining the Capacity of the Stomach and the Amount of Residual Contents." *Modern Medicine and Bacteriological Review* 5, no. 7 (1896): 171–175.

———. "Experimental Research Relating to Salivary Secretion and Digestion." In *Report of the Laboratory of Hygiene Connected with the Battle Creek Sanitarium,* reprinted from *Modern Medicine,* February and May 1895.

———. *Flaked Cereals and Process of Preparing Same.* Washington, D.C.: United States Patent Office, 1896.

———. *The Household Monitor of Health.* Battle Creek, Mich.: Good Health Publishing Co., 1891.

———. "The Ideal Sanitarium." Lecture read before the American Climatological Society, September 2, 1890.

———. "The Influence of an Aseptic Dietary upon the Bacteria of the Stomach." *Modern Medicine and Bacteriological Review* 4, no. 8 (1895): 215–217.

———. *The Itinerary of a Breakfast: A Popular Account of the Travels of a Breakfast through the Food Tube and of the Ten Gates and Several Stations through Which It Passes, Also of the Obstacles Which It Sometimes Meets.* New York: Funk and Wagnalls Company, 1918.

———. *The Natural Diet of Man.* Battle Creek, Mich.: The Modern Medicine Publishing Co., 1923.

———. "The New Chemistry of the Stomach." *The Bacteriological World and Modern Medicine* 1, no. 13 (1892): 430–435.

———. "A New Procedure for Suspension of the Pelvic Colon." *The Bulletin of the Battle Creek Sanitarium and Hospital Clinic* 18, no. 2 (1922): 80–88.

———. *Rational Hydrotherapy: A Manual of the Physiological and Therapeutic Effects of Hydriatic Procedures, and the Technique of Their Application in the Treatment of Disease.* Philadelphia: F.A. Davis Co., 1901.

———. "Relation of Modern Physiological Chemistry to Dietetics." *Modern Medicine and Bacteriological Review* 3, no. 4 (1894): 82–85.

———. "Relation of Modern Physiological Chemistry to Vegetarianism." *Modern Medicine and Bacteriological Review* 3, no. 3 (1894): 58–59.

———. "Sanitary Associations." *Seventh Annual Report of the Michigan State Board of Health* (1879): 83–95.

———. "Sanitarium Lectures: How to Live a Century," 1884. Manuscript held at the Bentley Historical Library, University of Michigan.

———. "Sanitarium Lectures: Nature's Method of Defending the Body against Disease," June 5, 1891. Manuscript held at the Bentley Historical Library, University of Michigan.

———. *The Stomach: Its Disorders and How to Cure Them*. Battle Creek, Mich.: Modern Medicine Publishing Co., 1896.

———. "A Successful and Dependable Method of Changing the Intestinal Flora." *Bulletin of the Battle Creek Sanitarium and Hospital Clinic* 22 (1928): 115–128.

———. "The Treatment of Hyperpepsia." *Modern Medicine and Bacteriological Review* 3, no. 4 (1894): 94–95.

———. *The Uses of Water in Health and Disease: A Practical Treatise on the Bath, Its History and Uses*. Battle Creek, Mich.: The Health Reformer, 1876.

Kirsch, Scott. "Cultural Geography III: Objects of Culture and Humanity, or, Re-'Thinging' the Anthropocene Landscape." *Progress in Human Geography* 39, no. 6 (2015): 818–826.

Kirsner, Joseph B. *The Development of American Gastroenterology*. New York: Raven Press, 1990.

Klingle, Matthew W. "Spaces of Consumption in Environmental History." *History and Theory* 42, no. 4 (2003): 94–110.

Kloppenburg, Jack Ralph. *First the Seed: The Political Economy of Plant Biotechnology, 1492–2000*. New York: Cambridge University Press, 1988.

Koch, Robert. *Essays of Robert Koch*. Translated by C. Carter. New York: Greenwood Press, 1987.

Kuhn, Thomas S. *The Structure of Scientific Revolutions*. Chicago: University of Chicago Press, 1962.

Langston, Nancy. *Toxic Bodies: Hormone Disruptors and the Legacy of DES*. New Haven, Conn.: Yale University Press, 2010.

Latour, Bruno. *Pandora's Hope: Essays on the Reality of Science Studies*. Cambridge, Mass.: Harvard University Press, 1999.

———. *Reassembling the Social: An Introduction to Actor-Network-Theory*. New York: Oxford University Press, 2005.

Lea, Jennifer. "Retreating to Nature: Rethinking 'Therapeutic Landscapes.'" *Area* 40, no. 1 (2008): 90–98.

Lears, T. J. Jackson. *No Place of Grace: Antimodernism and the Transformation of American Culture, 1880–1920*. New York: Pantheon, 1981.

Legan, Marshall Scott. "Hydropathy in America: A Nineteenth Century Panacea." *Bulletin of the History of Medicine* 45, no. 3 (1971): 267–280.

Liebig, Justus von. *Organic Chemistry in Its Applications to Agriculture and Physiology*. London: Taylor and Walton, 1840.

———. *Familiar Letters on Chemistry, and Its Relation to Commerce, Physiology and Agriculture*. Philadelphia: Campbell, 1843.

———. *Researches on the Chemistry of Food, and the Motion of the Juices in the Animal Body.* Lowell, Mass.: Daniel Bixby, 1848.

Livingstone, David. *Putting Science in Its Place: Geographies of Scientific Knowledge.* Chicago: University of Chicago Press, 2003.

Loughborough, J. N. "Report from Bro. Loughborough." *Review and Herald* 28 (September 1866).

Lowe, Bernice B. *Tales of Battle Creek.* Battle Creek, Mich.: Albert L. and Louise B. Miller Foundation, 1976.

Lowenthal, David. *The Past Is a Foreign Country.* New York: Cambridge, 1985.

Marcus, Alan I. *Agricultural Science and the Quest for Legitimacy: Farmers, Agricultural Colleges, and Experiment Stations, 1870–1890.* Ames: Iowa State University Press, 1985.

Martin, Emily. *Flexible Bodies: Tracking Immunity in American Culture from the Days of Polio to the Age of AIDS.* Boston: Beacon Press, 1994.

———. "Fluid Bodies, Managed Nature." In *Remaking Reality: Nature at the Millenium,* edited by B. Braun and N. Castree. New York: Routledge, 1998.

Marvin, Simon, and Will Medd. "Metabolisms of Obe*city:* Flows of Fat through Bodies, Cities, and Sewers." *Environment and Planning A* 38 (2006): 313–324.

Massie, Larry B., and Peter J. Schmitt. *Battle Creek: The Place behind the Products.* Woodland Hills, Calif.: Windsor Publications, 1984.

Matless, David. "Describing Landscape: Regional Sites." *Performance Research* 15, no. 4 (2010): 72–82.

McGerr, Michael. *A Fierce Discontent: The Rise and Fall of the Progressive Movement in America, 1870–1920.* New York: Free Press, 2003.

McKelvey, Hugh E. "Some History of Beautiful Goguac Lake." Manuscript held at the Willard Library, Helen Warner Branch, Battle Creek, Mich., ca. 1995.

Melosi, Martin V. *The Sanitary City: Urban Infrastructure in America from Colonial Times to the Present.* Baltimore, Md.: Johns Hopkins University Press, 2000.

Mennell, Stephen. *All Manners of Food: Eating and Taste in England and France from the Middle Ages to the Present.* New York: Blackwell, 1985.

Metcalfe, Richard. *Life of Vincent Priessnitz: Founder of Hydrotherapy.* London: Simpkin, Marshall, Hamilton, Kent and Co., 1898.

Michigan State Gazetteer and Business Directory. Vol. XXIV. Detroit: R. L. Polk and Co., 1905–1906.

Mintz, Sidney Wilfred. *Sweetness and Power: The Place of Sugar in Modern History.* New York: Penguin Books, 1986.

Mitman, Gregg. "In Search of Health: Landscape and Disease in American Environmental History." *Environmental History* 10, no. 2 (2005): 104–210.

———. *Breathing Space: How Allergies Shape Our Lives and Landscapes*. New Haven, Conn.: Yale University Press, 2007.

Money, John. *The Destroying Angel: Sex, Fitness and Food in the Legacy of Degeneracy Theory, Graham Crackers, Kellogg's Corn Flakes and American Health History*. Buffalo, N.Y.: Prometheus Books, 1985.

Murdoch, Jonathan. "The Spaces of Actor-Network Theory." *Geoforum* 29, no. 4 (1998): 357–374.

Nash, Linda. *Inescapable Ecologies: A History of Environment, Disease, and Knowledge*. Berkeley: University of California Press, 2006.

National Digestive Diseases Information Clearinghouse. *Irritable Bowel Syndrome*. Bethesda, Md.: National Institute of Diabetes and Digestive and Kidney Diseases, 2013.

Nissenbaum, Stephen. *Sex, Diet, and Debility in Jacksonian America: Sylvester Graham and Health Reform*. Westport, Conn.: Greenwood Press, 1980.

Numbers, Ronald L. *Prophetess of Health: Ellen G. White and the Origins of Seventh-day Adventist Health Reform*. Knoxville: University of Tennessee Press, 1992.

———. "Sex, Science, and Salvation: The Sexual Advice of Ellen G. White and John Harvey Kellogg." In *Right Living: An Anglo-American Tradition of Self-Help Medicine and Hygiene*, edited by C. E. Rosenberg. Baltimore, Md.: The Johns Hopkins University Press, 2003.

Nye, David, ed. *Technologies of Landscape: From Reaping to Recycling*. Amherst: University of Massachusetts Press, 1999.

Olmstead, Alan L., and Paul W. Rhode. *Creating Abundance: Biological Innovation and American Agricultural Development*. New York: Cambridge University Press, 2008.

"101 and We've Just Begun: Cereal Manufacturing Companies in Battle Creek." *Heritage Battle Creek: A Journal of Local History* 2 (Spring 1992): 51–56.

Order of the Common Council. *Charter of the City of Battle Creek, Michigan*. Battle Creek, Mich.: Gage and Sons, 1891.

Osborne, Thomas. "Security and Vitality: Drains, Liberalism and Power in the Nineteenth Century." In *Foucault and Political Reason: Liberalism, Neo-Liberalism and Rationalities of Government*, edited by A. Barry, T. Osborne, and N. Rose. Chicago: University of Chicago Press, 1996.

Parker, Alan (director). *The Road to Wellville*. United States: Columbia Pictures, 1994.

Pépin, Jacques. *La Technique: The Fundamental Techniques of Cooking: An Illustrated Guide*. New York: Pocket Books, 1978.

Pollan, Michael. "Our National Eating Disorder." *New York Times Magazine*, October 17, 2004.

———. *The Omnivore's Dilemma: A Natural History of Four Meals*. New York: Penguin, 2006.

Powell, Horace B. *The Original Has This Signature—W.K. Kellogg: The Story of a Pioneer in Industry and Philanthropy.* Battle Creek, Mich.: W.K. Kellogg Foundation, 1989. Originally published in 1956.

Priessnitz, Vincent. *The Cold Water Cure: Its Principles, Theory, and Practice with Ample Directions for Its Self-Application.* London: William Strange, 1843.

Prout, William. *Chemistry, Meteorology, and the Function of Digestion: Considered with Reference to Natural Theology.* London: W. Pickering, 1834.

Rabinbach, Anson. *The Human Motor: Energy, Fatigue, and Origins of Modernity.* Berkeley: University of California Press, 1992.

Rabinow, Paul, ed. *The Foucault Reader.* New York: Pantheon Books, 1984.

Rathbun, Earl Henry. "Millions in Every Bushel." Typed memoir, University of Michigan Bentley Historical Library, ca. 1920.

Rather, L.J. "The 'Six Things Non-Natural': A Note on the Origins and Fate of a Doctrine and a Phrase." *Clio Medica* 3 (1968): 337–347.

Riddell, F. "The Other Side of the Food Question." *Michigan Farmer and State Journal of Agriculture* 23, no. 29 (1892).

Rosenberg, Charles E. "Science, Technology, and Economic Growth: The Case of the Agricultural Experiment Station Scientist, 1875–1914." In *Nineteenth-Century American Science: A Reappraisal,* edited by G.H. Daniels. Evanston, Ill.: Northwestern University Press, 1972.

———. *The Care of Strangers: The Rise of America's Hospital System.* New York: Basic Books, 1987.

Rossiter, Margaret, W. *The Emergence of Agricultural Science: Justus Liebig and the Americans, 1840–1880.* New Haven, Conn.: Yale University Press, 1975.

Sack, Robert David. *Homo Geographicus: A Framework for Action, Awareness, and Moral Concern.* Baltimore, Md.: Johns Hopkins University Press, 1997.

Schneider, Daniel. *Hybrid Nature: Sewage Treatment and the Contradictions of the Industrial Ecosystem.* Cambridge, Mass.: MIT Press, 2011.

Schwartz, Norah Anita, Christine Alysse von Glascoe, Victor Torres, Lorena Ramos, and Claudia Soria-Delgado. "'Where They (Live, Work and) Spray': Pesticide Exposure, Childhood Asthma and Environmental Justice among Mexican-American Farmworkers." *Health and Place* 32 (2015): 83–92.

Schwarz, Richard. "John Harvey Kellogg: American Health Reformer." Dissertation, History, University of Michigan, Ann Arbor, 1965.

———. *John Harvey Kellogg, M.D.* Nashville, Tenn.: Southern Publishing Association, 1970.

———. *John Harvey Kellogg, M.D.: Pioneering Health Reformer.* Hagerstown, Md.: Review and Herald Pub. Association, 2006.

Schwieder, Dorothy. "Agricultural Issues in the Middle West, 1865–1910." In *Agriculture and National Development: Views on the Nineteenth Century*, edited by L. Ferleger. Ames: Iowa State University Press, 1990.

Scott, James C. *Seeing Like a State: How Certain Schemes to Improve the Human Condition Have Failed*. New Haven, Conn.: Yale University Press, 1998.

"Seed Wheat." *Michigan Farmer and State Journal of Agriculture* 23, no. 35 (1892).

Serres, Michel. *The Parasite*. Baltimore, Md.: Johns Hopkins University Press, 1982.

Shapiro, Laura. *Something from the Oven: Reinventing Dinner in 1950s America*. New York: Viking, 2004.

Shenstone, W. A. *Justus Von Liebig: His Life and Work (1803–1873)*. New York: MacMillan, 1895.

Sinclair, Upton. *The Jungle*. New York: Penguin Books, 1906.

Smith, Andrew F. *Eating History: 30 Turning Points in the Making of American Cuisine*. New York: Columbia University Press, 2009.

Smith, Erwin F. "Sewerage and Water-Supply: Discussion at a Sanitary Convention Held at Lansing, Mich." Reprinted from a supplement to the *Annual Report of the Michigan State Board of Health*, reprint no. 231 (1885): 104–106.

Smith-Howard, Kendra. *Pure and Modern Milk: An Environmental History since 1900*. New York: Oxford University Press, 2014.

Stage, Sarah. "Ellen Richards and the Social Significance of the Home Economics Movement." In *Rethinking Home Economics: Women and the History of a Profession*, edited by S. Stage and V. B. Vincenti. Ithaca, N.Y.: Cornell University Press, 1997.

Steingraber, Sandra. *Living Downstream: An Ecologist Looks at Cancer and the Environment*. New York: Addison-Wesley Publishing Co., 1997.

Storer, Norman W. *Science and Scientists in an Agricultural Research Organization: A Sociological Study*. New York: Arno Press, 1980.

Strauss, David. *Setting the Table for Julia Child: Gourmet Dining in America, 1934–1961*. Baltimore, Md.: Johns Hopkins University Press, 2011.

Straw, H. Thompson. "Battle Creek: A Study in Urban Geography." Unpublished Thesis, University of Michigan, Ann Arbor, Mich., 1938.

Swyngedouw, Erik. "Circulations and Metabolisms: (Hybrid) Natures and (Cyborg) Cities." In *Technonatures: Environments, Technologies, Spaces and Places in the Twenty-First Century*, edited by C. Wilbert and D. F. White. Waterloo, Ont.: Wilfrid Laurier University Press, 2009.

Tarr, Joel A. *The Search for the Ultimate Sink: Urban Pollution in Historical Perspective*. Akron, Ohio: University of Akron Press, 1996.

Tomes, Nancy. *The Gospel of Germs: Men, Women, and the Microbe in American Life*. Cambridge, Mass.: Harvard University Press, 1998.

———. "Spreading the Germ Theory: Sanitary Science and Home Economics, 1880–1930." In *Rethinking Home Economics: Women and the History of a Profession,* edited by S. Stage and V. B. Vincenti. Ithaca, N.Y.: Cornell University Press, 1997.

Trenev, Natasha. *Probiotics: Nature's Internal Healers.* Garden City Park, N.Y.: Avery, 1998.

True, A. C. *A History of Agricultural Experimentation and Research in the United States, 1607–1925.* Washington, D.C.: U.S. Department of Agriculture Miscellaneous Publication no. 251, Government Printing Office, 1937.

Turner, Frederick Jackson. "The Significance of the Frontier in American History." In *The Frontier in American History.* Malabar, Fla.: R. E. Krieger, 1975. Originally published in 1920.

Urla, Jacqueline, and Jennifer Terry. "Introduction: Mapping Embodied Deviance." In *Deviant Bodies: Critical Perspectives on Difference in Science and Popular Culture,* edited by J. Terry and J. Urla. Bloomington: Indiana University Press, 1995.

Valencius, Conevery Bolton. *The Health of the Country: How American Settlers Understood Themselves and Their Land.* New York: Perseus Books, 2002.

Veblen, Thorstein. "The Food Supply and the Price of Wheat." *The Journal of Political Economy* 1, no. 3 (1893): 365–379.

Waggoner, Paul E. "Research and Education in American Agriculture." *Agricultural History* 50, no. 1 (1976): 230–247.

Walker, Richard. *The Conquest of Bread: 150 years of Agribusiness in California.* New York: New Press, 2004.

Weiss, Harry B., and Howard R. Kemble. *The Great American Water-Cure Craze: A History of Hydropathy in the United States.* Trenton, N.J.: The Past Times Press, 1967.

White, Ellen. *Health, or, How to Live.* Battle Creek, Mich.: Steam Press of the Seventh-day Adventist Publishing Association, 1865.

Whorton, James. *Crusaders for Fitness: The History of American Health Reformers.* Princeton, N.J.: Princeton University Press, 1982.

———. *Inner Hygiene: Constipation and the Pursuit of Health in Modern Society.* New York: Oxford University Press, 2000.

Willits, Edwin. "Industrial Education." In Department of Agriculture, *Miscellaneous Special Report No. 9: Proceedings of a Convention of Delegates from Agricultural Colleges and Experiment Stations Held at the Department of Agriculture.* Washington, D.C.: Government Printing Office, 1885.

Wilson, Brian C. *Dr. John Harvey Kellogg and the Religion of Biologic Living.* Bloomington: Indiana University Press, 2014.

Winters, Donald L. "The Economics of Midwestern Agriculture, 1865–1900." In *Agriculture and National Development: Views on the Nineteenth Century,* edited by L. Ferleger. Ames: Iowa State University Press, 1990.

Worster, Donald. *Nature's Economy: A History of Ecological Ideas*. New York: Cambridge University Press, 1977.

Wylie, John. "Landscape and Phenomenology." In *The Routledge Companion to Landscape Studies*, edited by P. Howard, I. Thompson, and E. Waterton. New York: Routledge, 2013.

———. "A Single Day's Walking: Narrating Self and Landscape on the South West Coast Path." *Transactions of the Institute of British Geographers* 30 (2005): 234–247.

Youngman, S. J. "What Crop Will Pay?" *Michigan Farmer and State Journal of Agriculture* 23, no. 18 (1892).

Index

Page references followed by *fig.* indicate an illustration and *t* indicates a table. "JHK" refers to John Harvey Kellogg, "WKK" refers to William Keith Kellogg, and "the sanitarium" refers to the Battle Creek Sanitarium.

actor-network theory, 8–9, 18–19, 168
additives, 173
Advance Thresher (Battle Creek), 16, 145
Adventists. *See* Seventh-day Adventists
advertising, 156, 159–60
agricultural colleges, 135–36
agricultural experiment stations, 73, 75, 127, 133–36, 138, 148, 165
agricultural science, 15, 124–54; agro-industrial complex, 78; and appropriationism, 138, 141–42; Atwater's influence on JHK, 137–38, 139 *t;* Atwater's role in, 136–37; bottleneck effect in farming, 144; and economic difficulties/discontent of farmers, 127–28, 140, 158; extensification vs. intensification in farming, 142–44; farming vs. mechanic arts, 140–41; food chemistry, 124–25, 129–30, 130 *fig.*; German, 130–33, 194n22; government-funded research in, 129–30; grain-threshing machines, 16; industrialization of farming, 128, 141–42; legislation in support of, 135; Midwestern (19th cent.), 126–28; and nutritional science, 125, 130, 136–37; opposition to, 147; overview of, 124–26; political economy of, 100; seed companies, 124, 149–50, 151–53 *figs.*, 153–54; systematization of farming, 128, 141, 144; threshers used in, 144–48, 146 *figs.*; U.S. and southern Michigan, 133–36
Alaimo, Stacy, 2–3
Alcott, William, 20, 25, 28, 83
alimentary canal, 117 *fig.*, 129
alternative food movements, 172–74
American Medical Association (AMA), 18, 157
American Pure Food Company (Battle Creek), 155–56
American Seed Trade Association, 149
American Vegetarian Society, 26, 42
anemia, 129
antimodernism, 164–67
apepsia, 55, 59 *fig.*, 70
Atwater, Wilbur: *The Chemical Composition of American Food Materials,* 138; on chemistry's role in agricultural intensification, 143; dietary charts of, 11, 74–75; JHK influenced by, 137–38, 139 *t;* Liebig's

Atwater, Wilbur *(continued)*
influence on, 73; nutritional guides by, 137; nutrition research by, 73–74; on refuse and ash, 138; role in agricultural science, 135–37
auto-intoxication: Bouchard on, 55, 114–15; and conversion of grains, 84; diseases caused by, 12; efficient waste removal to avoid, 15; germs' role in, 53–54, 60, 114; JHK adopts theory of, 55, 76, 84, 85 *fig.*, 114; JHK cures, 114–16, 117–18 *figs.*, 119; stagnation of food as cause of, 114

bacteria/germs: as bred by food, 53–54; discovery of, 114; disease caused by, 53–54, 60, 114 (*see also under* typhoid fever); fermentation accompanied by, 55; germ theory, 47, 53, 109–10; infections caused by, 114–15; JHK on, 15, 44, 53–54, 64, 76, 113, 184n43; probiotics, 12, 100, 173–74; in sewage/human waste, 14–15; soil as a filter for, 108
bacteriology, 39, 53, 62, 68, 108–11, 113
Battle Creek (Michigan): as Cereal City, 156; factories in, 155; growth of, 106; health food companies in, 12, 155–56; industrial growth in, 191n20; population of, 106, 191n20; restaurants in, 71; sewer system in, 14–15, 102, 105 *fig.*, 106, 107; water for, 119, 121, 121 *fig.*
Battle Creek Board of Public Works, 102
Battle Creek Breakfast Food Company, 155–56
Battle Creek Cereal Food Company, 155–56
Battle Creek Flaked Food Company, 155–56
Battle Creek Food Company, 155–56
Battle Creek Food Products Company, 155–56
Battle Creek Pure Food Company, 155–56
Battle Creek Sanitarium (*formerly* Western Health Reform Institute), 20–45; advertisements for, 44; antimodernism at, 164–65; chemical laboratories at, 42; daily routines for patients at, 26; diet at, 25, 40, 57, 71–72, 99, 150, 151 *fig.*; Diet Service Department, 116; evacuation schedule for patients, 14; examination room, 49, 50 *fig.*; exercise regimens at, 1; expansion of, 44–45; experimental kitchens at, 14, 42 (*see also under* flaked cereal; stomach); finances of, 30–31, 33, 38–39, 62; foods originating at, 21, 62–64 (*see also* corn flakes; granola); graphic method of lab analysis used at, 58, 59 *fig.*, 60; greenhouses, farms, and mill of, 150, 154; health foods invented/perfected at, 1; health reform promoted at, 99; hydrotherapy at, 25–29, 38–43, 75, 119; intake procedure at, 49; JHK as director of, 11–12, 31–33, 37–39, 75, 112–13; leadership of, early, 25, 30; lectures at, 1; lifestyle promoted at, 34; modernization of, 164–65; name change/rebranding as a sanitarium, 37–38; natural cure vs. medical science at, 25, 33–34, 42, 44; opening of, 24–25; philosophy of nutritional health at, 1, 33; as place, 37–38; as a place of healing, 22–26, 44; portrayals of, 181n49; secularization of, 63; spiritual advantages espoused at, 31 (*see also* Seventh-day Adventists); technologies/machinery used at, 1, 3, 14, 40, 41 *fig.* (*see also* rollers, cereal); therapeutic landscape around, 34–36; treatments/machinery at, 25–26; water for, 119, 121, 121 *fig.*
Battle Creek Sanitarium Health Food Company, 84, 85 *fig.*, 86–87, 187n22
Battle Creek Toasted Corn Flake Company, 155–57. *See also* Sanitas Nut Food Company
Beaumont, William, 182n14
Bellevue Hospital Medical College (New York City), 32–33
Bemis, Edward, 140
biopower, 10, 34, 137
body-environment relationships: body as part of an ecological system, 100; environment as exoskeleton for the body, 171, 199n50; food-body, 10–11, 82, 84, 100–101; landscape-body, 8–9, 103, 123, 170–71; machine-body, 90–94, 94 *fig.*, 101; overview of, 3, 168–69
Bouchard, Charles, 53, 55, 63, 114–15, 191n25
Boyle, T. Coraghessan: *The Road to Wellville*, 181n49
bran bread, 82–83
bread making, 82, 86–87, 186n13
breakfast cereal. *See* cereal
brewer's yeast, 81–82
Broussais, François, 82
Butler, Edith, 145, 147

California, fruit and vegetable industry in, 149
calorie counting, 66, 100, 139 *t*
"cancer alley" (Louisiana), 103
capitalist agricultural economy, 159

carbohydrates: avoidance of, 66; calculation of, 100, 138; on Diet Lists, 70–71; recommended daily amount of, 67 *fig.*, 139 *t;* required for bodily activity, 137
Carson, Gerald, 192n2
Carson, Rachel: *Silent Spring*, 7
catarrh, 69
celiac disease, 173
cereal: advertising for, 156, 159–60; and the Battle Creek name, 160–61; decision to use corn, 157–59; and the food processing revolution, 161–62; with milk, 162–63, 177n20; overview of, 155–57; sugar in, 159. *See also* corn flakes; flaked cereal; wheat flakes
cesspools, 108, 113
The Chemical Composition of American Food Materials (Atwater), 138
Chemistry, Meteorology, and the Function of Digestion (Prout), 162
The Chemistry of Cooking and Cleaning (Richards), 68
chemistry training in Germany vs. the United States, 73–74
chemists at agricultural experiment stations, 134–35
Chicago Board of Trade, 158–59
chlorine, 49, 55, 129, 182n12
cholera, 15, 53–54, 61, 104, 111, 123
Cincinnati, 158
Claridge, R. T., 36, 41
Clark, Arnold, 107–8
Cochrane, Willard, 144
colonic machines, 119, 120 *fig.*, 121, 122 *fig.*
Connecticut agricultural experiment station, 133, 135
conservation movement, 131
constipation, 69
Coole, Diana, 17
corn, 97, 157–59
corn flakes: ads for, 160; decision to use corn, 157–59; Granose (wheat flakes) as precursor to, 71, 94 *fig.*; as a health food, 1, 157; invention of, 1, 21, 71, 100–101
corn syrup, 78
countercultural food movements, 78–79
Crohn's disease, 4, 173
cuisine: American, 12, 17, 78–79, 81, 94–98, 100–101; dangers in the prevailing food system, 78–79; geographical component of, 95; health reform, 79, 83, 97–99, 187n17 (*see also* health reform movement); vs. kitchen, 95–96, 189n43; manufactured/engineered foods in, 77–78, 99; meaning of, 94–95; national cuisines, 97; overstimulating, 82

Daily Inter Ocean, 182n9
Davy, Humphry, 131–32
defecation schedule, 14, 116, 119
"Detroit" (Konner), 181n49
deviance, signs of, 61
dextrin, 13, 49, 84–85, 89, 91, 93
diabetes, 69
diagnostic charts/standardization, 58, 59 *fig.*, 60–61, 65 fig., 66, 69, 74–75
Diamond, Harvey and Marilyn: *Fit For Life*, 54
diarrhea, 173
diet: Adventists' promotion of (*see under* Seventh-day Adventists); American obsession with eating right, 42–43; for the American worker, 74; anxieties about, 66; Atwater's charts for, 11, 74–75; diet prescriptions/prescription meals, 11–12, 42–43, 48, 63–64, 65 *fig.*, 72, 129, 138 (*see also* Granose); disease linked with, 42–43, 47, 98–99; fad diets, 54–55; JHK's charts for/research on, 11, 42, 138, 139 *t*, 164; meat eliminated from, 116; national, 125; natural antibiotic based on, 54; at the sanitarium, 25, 40, 57, 71–72, 99, 150, 151 *fig.*; sterilized food, 57, 62–63; and water cures, 40–42. *See also* cuisine; stomach
Diet Lists, 69–71
digestive reductionism, 1
digestive system: as biotechnology, 16; debates about health of, 173; diagrams of, 116, 117–18 *figs.*; disorders of (*see* apepsia; dyspepsia; hyperpepsia/hypopepsia); as an ecosystem, 173–74; environmental geography for, 4, 75; extensibility of, 103–4; health's dependence on, 4, 110; in isolation, study of, 43, 46–47; quality/healthfulness of foods diagnosed via, 4; role in the landscape-body relationship, 8–9, 103, 123, 170–71; stagnation of food in, 54–55, 72, 114–16, 123; technologies of, 2, 4, 13–19 (*see also* rollers, cereal); time table of, 118 *fig. See also* stomach; urban landscape as extension of the digestive system
diphtheria, 162–63
DuPuis, Melanie, 169
dyspepsia, 5, 47–48, 55. *See also* indigestion

Electric Bed, 40, 41*fig.*
electricity, 40
electric shock, 26
Emerick, C. F., 143
Emerson, Ralph Waldo, 106
enemas, 115–16, 119, 165. *See also* colonic machines
Enteria, 169
environmentalism, 7, 173–74

Farmer's Alliance, 127
farming. *See* agricultural science
fats: avoidance of, 66; calculation of, 100, 138; on Diet Lists, 70–71; recommended daily amount of, 67*fig.*, 139 *t;* required for bodily activity, 137
fatty acids, 49
FDA (Food and Drug Administration), 67*fig.*
fecal bacteriotherapy, 4
fecal-oral diseases, 104, 107, 121. *See also* cholera; typhoid fever
feminists on signs of deviance, 61
fertilizers, 143
fever, 69
Fit For Life (the Diamonds), 54
flailing, 144–45
flaked cereal, 14, 77–101; consumer base for, 173; experiment kitchen's role in, 79, 96–97; Sylvester Graham's role in, 81–83; vs. granola, 79–80, 83–85, 85*fig.*, 100, 187n22; Granose, 62, 65*fig.*, 68–69, 71, 79, 89, 93, 94*fig.*; invention of, 77, 86–89, 187–88n27; and the machine-body relationship, 90–94, 94*fig.*, 101; marketing-rights dispute over, 187n27; overview of, 77–81; patent on, 90–91; popularity of, 89–90. *See also* corn flakes; rollers, cereal
flour, 81–82, 86
food combining, 54
Food Guide Pyramid, 137
food labels, 67*fig.*
Food Materials and Their Adulterations (Richards), 68
food movements, 78–79, 172–74
food processing revolution, 161–62
food safety, trust in, 173
food studies, 8–9, 11, 169
forestry schools, German, 130–31
Foucault, Michel, 10, 34, 68, 137
Frost, Samantha, 17

Gandy, Matthew, 199n50
general practitioners, 48
geographical origins of food, 2
geography of science, 24
geological survey, 133
Germany: agricultural science in, 130–33, 194n22; universities in, 73–74
germs. *See* bacteria/germs
Gerstner, Patsy, 150
Gesler, Wilbert M., 178n6
Ginley, J. H., 30
gluten intolerance, 173
Goguac Lake (Michigan), 119, 121, 121*fig.*
Good Health, 108
gout, 69
Graefenberg. *See* Jesenik
Graham, Robert Hay, 180n40
Graham, Sylvester: bran bread promoted by, 82–83; on the food-body relationship, 82, 100; health foods invented by, 83, 86, 92; as health reformer, 17, 20, 25, 81, 92, 99; influence of, 28, 80; moral/social eating philosophy of, 101, 172; on overstimulation, 82–83; on refined flour, 81–82; role in flaked cereal, 81–83; on technology as corrupting foods, 99
Grange, 127
granola, 79–80, 83–85, 85*fig.*, 100, 187n22
Granose (wheat flakes), 62, 65*fig.*, 68–69, 71, 79, 89, 93–94, 94*fig.*
Great Plains, 158

Hammond (Harry) Seedsman (Fifield, Mich.), 150, 151–53*figs.*, 153–54
harvesters, 138, 140
Hatch Act (1887), 135, 148
Hayles, Katherine, 16
healing, places of, 22–26, 44, 68
health foods, manufactured: American cuisine influenced by, 12–13; back-to-nature vs. to-the-future foods, 99–100; vs. home cooking, 12; vs. organic farming, 12–13; rise of, 148; stomach health as basis of (*see* stomach). *See also* Granose
The Health Reformer, 24, 30, 32, 44
health reform movement: bland foods from, 17, 78, 83, 187n17; cuisine of, 79, 83, 97–99, 187n17; culinary arm of, 86–87, 99; dictates of, 20; drugs forbidden by, 42; food philosophy of, 92; Sylvester Graham's role in, 17, 20, 25, 81–82, 92, 99; JHK associated with, 17–18, 99, 172; and religion, 20–22, 28–29, 33; temperance aspect of, 31; and the Vegetarian Society, 42

health vacations, 44. *See also* spa towns; water cures
Hetch Hetchy Valley (Yosemite National Park), 131
home economics, 68
Homestead Act (1862), 142
household economy, commercialization of, 82
How Crops Feed (Johnson), 136
How Crops Grow (Johnson), 136
human waste: bacteria in, 14–15; movement of, 113, 123 (*see also* sewers); pathway of, 11
Humboldt, Alexander von, 132–33
humus, 131–32
husbandry, 158
hydrochloric acid, 46, 48, 59 *fig.*, 60, 71, 129
hydrotherapy (water cure) institutions, 26–29, 38–43, 79–80. *See also* Battle Creek Sanitarium
hygienic philosophy, 166–67
Hygieo-Therapeutic College (Florence Heights, N.J.), 32
hyperpepsia/hypopepsia, 55, 59 *fig.*, 60, 69–70, 129, 130 *fig.*

indigestion, 4, 14, 43, 47–49, 51 *fig.*, 126
industrialization, 106
intestinal infections, 173
irritable bowel syndrome, 4, 173

Jackson, James C.: on the American diet, 98; American Vegetarian Society founded by, 26; on diet and disease, 42, 98; granula invented by, 80, 83, 86; as health reformer, 17, 20; influence of, 28, 75, 79–80; water cure of, 26–28, 40, 44, 79
Jesenik (*formerly* Graefenberg, Czech Republic), 27–28, 36–37, 40–41, 179n19, 180n40
Johnson, Samuel, 135, 143; *How Crops Feed*, 136; *How Crops Grow*, 136
The Jungle (Sinclair), 161

Kalamazoo River, 102, 105, 121
Kautsky, Karl, 132
Kellogg, Ann (JHK's mother), 29, 31
Kellogg, Ella Eaton (JHK's wife): accomplishments of, 68; cooking schools attended by, 68; experimental kitchen of, 47, 66, 68–69, 79, 181n1; Granola Mush recipe of, 79, 83; on housekeeping/ventilation, 109; on nutritional science, 72; *Science in the Kitchen*, 66, 79, 95–96
Kellogg, John Harvey, 125; as an Adventist, 75; on advertising, 156; AMA's pressure on, 157; on American eating habits, 155; Atwater's influence on, 137–38, 139 *t;* auto-intoxication cured by, 114–16, 117–18 *figs.*, 119; on bacteria/germs, 15, 44, 53–54, 64, 76, 113, 184n43; on bacteriology, 39, 53–54, 62; biography of, 188n27; body's nature mistrusted by, 17; Bouchard's influence on, 114, 191n25; diagnoses linked to digestive disorders by, 47–48; diagnostic charts/standardization used by, 58, 59 *fig.*, 60–61, 65 fig., 66, 69, 74–75; dietary charts of/research by, 11, 42, 138, 139 *t*, 164; on digestive disorders, 3; as director of the sanitarium (*see under* Battle Creek Sanitarium); enemas developed by, 115, 119; flaked cereal invented by, 4–5, 13–14, 86 (*see also* corn flakes; wheat flakes); on the food-body relationship, 10–11, 84, 100–101; on food chemistry and physiology, 129; on food gates, 116, 117 *fig.*; food innovations by, 68–69, 80–81, 87 (*see also* flaked cereal); on foods shipped to the sanitarium, 150; food used as medicine by, 11–12, 42–43, 48, 63–64, 72, 129, 138 (*see also* Granose); as a gastroenterologist, 48, 112–13, 182n4; as generalist and specialist, 182n5; on granola, 80, 87; health program of, generally, 110, 191n20; as health reformer, 92, 99, 172; hydrotherapy approach of, 39–43, 75; medical philosophy of, 165–67; medical training of, 11, 15, 32–33, 43; on the Michigan State Board of Health, 15, 112–13; microscope used by, 75; on milk, 163; modernizing practices of, 164–65, 172; on the "modern stomach," 2, 125–26, 174; on a national diet, 125; philosophy of health, 1, 33–34; on Priessnitz, 39–40; on privies, 112; rigid diet regime of, 63; on the sanitarium's building and site, 35, 38; Sanitas Nut Food Company opened by, 187n22; on sanitation, 112; sincerity of his desire to help, 183–84n34; on starch, 84; *The Stomach*, 1, 3, 46, 66, 75; on stomach acids, 184n43; on the stomach's centrality to the human body, 46–47 (*see also* stomach); on sugar in cereal, 159; on sweet corn, 97; on technology as improving on nature, 92; as traditionalist and modernist, 33–34, 44–45; *The Uses of Water in Health and Disease*, 40; water cure

Kellogg, John Harvey *(continued)*
philosophy abandoned by, 39–43; on wheat flakes, 13; WKK's dispute with, 157, 159, 184n34, 187–88n27; writings of, 9–10
Kellogg, John Preston (JHK's father), 29–32
Kellogg, William Keith: advertising expenditures of, 160; on branding the Kellogg name, 157; flakes cereal perfected by, 90, 91 *fig.* (*see also* flaked cereal); food innovations by, 87; on freshness of foods, 162; JHK's dispute with, 157, 159, 184n34, 187–88n27; machinery invented by, 162; sanitarium's food company controlled by, 156–57, 159–60; signature on cereal boxes, 1, 161
Kellogg's: Battle Creek name's importance to, 160–61; food processing innovations at, 161–62. *See also* corn flakes
kidneys, 116
Koch, Robert, 104
Konner, Jeremy: "Detroit," 181n49
Kuhn, Thomas, 86

Laboratory of Hygiene, 49, 51–53, 52 *fig.*, 57, 58–59 *figs.*
landscapes: the body's relationship to, 8–9, 103, 123, 170–71; as epistemology, 5–7, 9, 15; moving through, 167–71; rural, reshaping of, 127; size/importance of, 169–70; therapeutic, 34–37; urban, reform of, 110. *See also* urban landscape as extension of the digestive system
Lane, Arbuthnot, 53
Latour, Bruno, 153, 168–69
Lay, Horatio S., 25–26, 30
Lears, Jackson, 164
Lehreiheit (freedom of teaching), 73–74
Lernfreiheit (freedom of learning), 73–74
Liebig, Justus von, 73, 131, 133, 137, 165; *Organic Chemistry*, 132, 134
Light Bath, 40
Lindsay, Harmon, 30
liver, 116

Machine, Vibrator Threshing, 145
maltose, 49
Mann, Horace, 20
maps, 105, 105 *fig.*
Marcus, Alan, 134
materialism, 17
McCutcheon, John T., 156
meat, demand for, 157
Meat Inspection Act (1906), 161
Mechanical Gym, 40
medical science: deviance denoted in, 61; vs. natural cure, 25, 33–34, 42, 44; philosophies of, 165–67; specialization in, 48, 166, 182n5
metabolism, 126
Metchnikoff, Elie, 53, 150
miasmatic theories of disease, 109–11, 113. *See also* putrefaction
Michigan State University (East Lansing), 133
microbes. *See* bacteria/germs
middle class, 61
migration, rural-to-urban, 145
milk, 162–63, 173, 177n20
Mintz, Sidney, 159
Morrill Land Grant Act (1862), 135
Morton, J. Sterling, 149

Nichols, John, 145
Nichols, Thomas Low, 99
Nichols & Shepard (*later named* Oliver Corporation; Battle Creek), 16, 145, 146 *figs.*, 147, 196n64
Nissenbaum, Stephen, 82, 186n15
nitrogen cycle, 132
noise/static, 167–68
nonnaturals, 165–67
nutritional science: and agricultural science, 125, 130, 136–37; birth of, 101, 130; nutrition labels, 67 *fig.*; and the stomach, 72–75, 125
Nuttose and Nuttolene (nut-based bars and spreads), 65 *fig.*

object-oriented philosophy, 7–8
Oliver Corporation. *See* Nichols & Shepard
Organic Chemistry (Liebig), 132, 134
organic farming, 12–13
Our Home on the Hillside (Dansville, N.Y.), 26, 42, 79–80
outhouses (privies), 102, 107–8, 112, 114
overstimulation, 80–83

packaging, 161–62, 173
Parker, Alan: *The Road to Wellville*, 181n49
pasteurization, 163, 173
Perky, Henry, 87
Pinchot, Gifford, 130–31
place: meaning of, 95; movement between places, 167; the sanitarium as, 37–38
plows, 138, 140, 142
pork, 157–58
Post, Charles, 156
posthumanism research, 17

Powell, Horace, 88, 156, 162
prairies, cultivation of, 142–43
preservatives, 78, 173
Priessnitz, Vincent, 39–40, 44, 179n19. *See also* Jesenik
Priessnitz Medical Spa (Jesenik, Czech Republic), 179n23
privies (outhouses), 102, 107–8, 112, 114
probiotics, 12, 100, 173–74
Progressives, 61
proteids, 49, 66, 67*fig.*, 70–71
protein, 47, 72–73, 100, 136–38, 182n12
Protestant millennialists, 20. *See also* Seventh-day Adventists
Prout, William: *Chemistry, Meteorology, and the Function of Digestion,* 162
public health: and dietary anxieties, 66; explaining the roots of diseases, 110–11; governmental programs, 106–7, 191n13; milk campaign, 163; patients' anxiety about improving, 61–62
Pure Food and Drug Act (1906), 161
putrefaction, 12, 47, 55, 109, 116

radio advertising, 160
railroads, 142, 157–58
Reeves, George, 160
Review and Herald, 30, 153
rheumatism, 69
Richards, Ellen: *The Chemistry of Cooking and Cleaning,* 68; *Food Materials and Their Adulterations,* 68
The Road to Wellville (book; Boyle), 181n49
The Road to Wellville (film; Parker), 181n49
rollers, cereal, 4–5, 13, 15, 87–93, 91*fig.*, 161–62
Rossiter, Margaret W., 132
rural economic distress, 127–28
Russell, William, 30

sanatoriums vs. sanitariums, 37
Sanitas Nut Food Company (*later named* Battle Creek Toasted Corn Flake Company), 14, 86–87, 93, 155–57, 187n22
sanitation, 68, 103, 107, 110, 112, 123, 162–63
Schwarz, Richard, 88, 182n4, 187–88n27
science. *See* agricultural science; medical science; nutritional science
Science in the Kitchen (E. E. Kellogg), 66, 79, 95–96
seeds: collecting/distributing, 148; reusing varieties of, 148–49; seed companies, 124, 149–50, 151–53*figs.*, 153–54
self-poisoning. *See* auto-intoxication
Serres, Michel, 167
Seventh-day Adventists: Battle Creek headquarters of, 30; flaked cereal's importance to, 187–88n27; health/diet promoted by, 21–22, 28–29, 31, 34, 39, 63, 75; inauguration of, 20; mission of, 30–31; printing press of, 29–30; spiritual message of, 28–29. See also Battle Creek Sanitarium; *The Health Reformer*
sewers: in Battle Creek, 14–15, 102, 104–5; complacency about/opposition to, 108, 110; for crowded cities, 107, 111; disease alleviated by, 104; European, 14, 107; as extension of digestive system, 102–4, 113; fecal-oral diseases from lack of, 104, 107; harmful bacteria in sewage, 14–15, 111, 113; history of, 14, 103–5, 105*fig.*; public health reformers on, 111–12
Shenstone, W. A., 131
Shepard, David, 145
Shew, Joel, 20
Shredded Wheat factory (Denver), 87
Silent Spring (Carson), 7
Sinclair, Upton: *The Jungle,* 161
Smith, Andrew, 81
Smith, Erwin, 111
soil, bacteria filtered by, 108
soil physics, 136
spa towns, 22–24, 178n6
stagnation, 54–55, 72, 114–16, 123
starch: conversion into dextrin, 13, 84–85, 89, 91–93, 94*fig.*; cooking of, 92, 188n36; testing for, 49
St. Martin, Alexis, 182n14
The Stomach (JHK), 1, 3, 46, 66, 75
stomach, 46–76; bacteria in, 53–54, 57, 58 *fig.*, 64 (*see also* auto-intoxication); and the experiment kitchen, 46, 57, 62–72, 65*fig.*, 67*fig.*, 181n1; fermentation in, 55, 71; food innovations based on chemistry of, 62–64 (*see also* flaked cereal); gastric juice (hydrochloric acid) in, 46, 48, 59 *fig.*, 60, 71, 129; health regime centered on, 46–62, 50–51*figs.*, 56*fig.*, 58–59 *figs.*; Laboratory of Hygiene used to study, 49, 51–53, 52*fig.*, 57, 58–59*figs.*; modern, 2, 125–26, 174; and nutritional science, 72–75, 125; peering into, 51–52, 182n14; testing contents of, 49, 50–51 *figs.*, 51–53, 55–60, 56*fig.*, 58–59*figs.*, 183n16
Straw, H. Thompson, 166

subject/object duality, 170
sugar, 159
sun baths, 26
supplements, 173
"Sweetheart of the Corn" advertisement, 160
Swyngedouw, Erik, 126

television advertising, 160
tempering, 88–90, 93
Thaer, Albrecht, 194n22
Thoreau, Henry David, 106
threshers, 16, 144–48, 146 *figs.*
Toasted Corn Flake Company, 160–61
Tocqueville, Alexis de, 131
toilets, flush, 103
toxicity, 7–8, 120 *fig.*
tractors, 138, 140
Trall, Russell T., 20, 32, 40, 44
transcendentalism, 106
typhoid fever: bacteria's role in, 14–15, 39, 53–54, 104, 111, 162–63; death toll from, 54; as a public health problem, 61, 98, 123; transmission of, 104, 162–63; and waste management, 107

University of Michigan, 32
urban decay, 62
urban industrialization, 35, 105–7, 105 *fig.*
urban landscape as extension of the digestive system, 14–15, 102–23; bacteriological theories of disease, 109–11 (*see also* bacteria/germs; bacteriology); Battle Creek as illustrating, 105–7, 105 *fig.*; colonic machines, 119, 120 *fig.*, 121, 122 *fig.*; extensibility of the digestive system, 103–4, 118 *fig.*; JHK cures auto-intoxication, 114–16, 117–18 *figs.*, 119; JHK's central role in, 112–13; miasmatic theories of disease, 109–11, 113 (*see also* putrefaction); overview of, 167–68; privies (outhouses), 102, 107–8, 112, 114; the stomach, without sewers, 104–5, 105 *fig.*
urban landscape reform, 110
USDA (U.S. Department of Agriculture), 127–28, 135–37, 148. *See also* agricultural experiment stations
U.S. Department of Agriculture. *See* USDA
The Uses of Water in Health and Disease (JHK), 40
U.S. Food and Drug Administration, 67 *fig.*

vacuum treatments, 26
vaults, waste-disposal, 108
Veblen, Thorstein, 143
ventilation, 68, 109
Vibrator Threshing Machine, 145, 146 *fig.*, 147–48
Vibro-Therapy Bed, 40
Voit, Carl von, 136

wages, 74
Walker, Richard, 140
water cures: and diet, 40–42; expected behavior at, 44, 181n54; as health vacations, 44; history of, 27–28; at hydrotherapy institutions, 26–29, 38–43, 79–80 (*see also* Battle Creek Sanitarium; Jesenik); by Jackson, 26–27; JHK's abandonment of, 39–43; vs. medicine, 26–27; minerals' healing powers, 23–24; at spa towns, 22–24, 178n6
Waxtite packaging, 161–62, 173
wells, contamination of, 108, 111
Western Health Reform Institute. *See* Battle Creek Sanitarium
wheat, 150, 157–58, 192n2
wheat flakes, 13. *See also* Granose
White, Ellen G.: Adventist doctors supported by, 32; on bacteriology, 39; diet regime of, 63; fundraising by, 29; health visions of, 20–22, 25, 28; influences on, 28; at Jackson's water cure, 28; JHK supported by, 32–33; on Lay, 30; on medical science, 32; on a place of healing, 22, 25–26; religious mission of, 30–31, 38; writings of, 24 (see also *The Health Reformer*). *See also* Battle Creek Sanitarium; Seventh-day Adventists
White, James, 29–33, 38–39
white bread, 81–82
Whole Foods, 93
Whorton, James, 82, 165–66
Wiley, Harvey, 127–28
Willits, Edwin, 140–41
Wonder Bread, 188n36

CALIFORNIA STUDIES IN FOOD AND CULTURE

Darra Goldstein, Editor

1. *Dangerous Tastes: The Story of Spices,* by Andrew Dalby
2. *Eating Right in the Renaissance,* by Ken Albala
3. *Food Politics: How the Food Industry Influences Nutrition and Health,* by Marion Nestle
4. *Camembert: A National Myth,* by Pierre Boisard
5. *Safe Food: The Politics of Food Safety,* by Marion Nestle
6. *Eating Apes,* by Dale Peterson
7. *Revolution at the Table: The Transformation of the American Diet,* by Harvey Levenstein
8. *Paradox of Plenty: A Social History of Eating in Modern America,* by Harvey Levenstein
9. *Encarnación's Kitchen: Mexican Recipes from Nineteenth-Century California: Selections from Encarnación Pinedo's* El cocinero español, by Encarnación Pinedo, edited and translated by Dan Strehl, with an essay by Victor Valle
10. *Zinfandel: A History of a Grape and Its Wine,* by Charles L. Sullivan, with a foreword by Paul Draper
11. *Tsukiji: The Fish Market at the Center of the World,* by Theodore C. Bestor
12. *Born Again Bodies: Flesh and Spirit in American Christianity,* by R. Marie Griffith
13. *Our Overweight Children: What Parents, Schools, and Communities Can Do to Control the Fatness Epidemic,* by Sharron Dalton
14. *The Art of Cooking: The First Modern Cookery Book,* by the Eminent Maestro Martino of Como, edited and with an introduction by Luigi Ballerini, translated and annotated by Jeremy Parzen, and with fifty modernized recipes by Stefania Barzini
15. *The Queen of Fats: Why Omega-3s Were Removed from the Western Diet and What We Can Do to Replace Them,* by Susan Allport
16. *Meals to Come: A History of the Future of Food,* by Warren Belasco
17. *The Spice Route: A History,* by John Keay

18. *Medieval Cuisine of the Islamic World: A Concise History with 174 Recipes,* by Lilia Zaouali, translated by M. B. DeBevoise, with a foreword by Charles Perry
19. *Arranging the Meal: A History of Table Service in France,* by Jean-Louis Flandrin, translated by Julie E. Johnson, with Sylvie and Antonio Roder; with a foreword to the English-language edition by Beatrice Fink
20. *The Taste of Place: A Cultural Journey into Terroir,* by Amy B. Trubek
21. *Food: The History of Taste,* edited by Paul Freedman
22. *M. F. K. Fisher among the Pots and Pans: Celebrating Her Kitchens,* by Joan Reardon, with a foreword by Amanda Hesser
23. *Cooking: The Quintessential Art,* by Hervé This and Pierre Gagnaire, translated by M. B. DeBevoise
24. *Perfection Salad: Women and Cooking at the Turn of the Century,* by Laura Shapiro
25. *Of Sugar and Snow: A History of Ice Cream Making,* by Jeri Quinzio
26. *Encyclopedia of Pasta,* by Oretta Zanini De Vita, translated by Maureen B. Fant, with a foreword by Carol Field
27. *Tastes and Temptations: Food and Art in Renaissance Italy,* by John Varriano
28. *Free for All: Fixing School Food in America,* by Janet Poppendieck
29. *Breaking Bread: Recipes and Stories from Immigrant Kitchens,* by Lynne Christy Anderson, with a foreword by Corby Kummer
30. *Culinary Ephemera: An Illustrated History,* by William Woys Weaver
31. *Eating Mud Crabs in Kandahar: Stories of Food during Wartime by the World's Leading Correspondents,* edited by Matt McAllester
32. *Weighing In: Obesity, Food Justice, and the Limits of Capitalism,* by Julie Guthman
33. *Why Calories Count: From Science to Politics,* by Marion Nestle and Malden Nesheim
34. *Curried Cultures: Globalization, Food, and South Asia,* edited by Krishnendu Ray and Tulasi Srinivas
35. *The Cookbook Library: Four Centuries of the Cooks, Writers, and Recipes That Made the Modern Cookbook,* by Anne Willan, with Mark Cherniavsky and Kyri Claflin

36. *Coffee Life in Japan,* by Merry White
37. *American Tuna: The Rise and Fall of an Improbable Food,* by Andrew F. Smith
38. *A Feast of Weeds: A Literary Guide to Foraging and Cooking Wild Edible Plants,* by Luigi Ballerini, translated by Gianpiero W. Doebler, with recipes by Ada De Santis and illustrations by Giuliano Della Casa
39. *The Philosophy of Food,* by David M. Kaplan
40. *Beyond Hummus and Falafel: Social and Political Aspects of Palestinian Food in Israel,* by Liora Gvion, translated by David Wesley and Elana Wesley
41. *The Life of Cheese: Crafting Food and Value in America,* by Heather Paxson
42. *Popes, Peasants, and Shepherds: Recipes and Lore from Rome and Lazio,* by Oretta Zanini De Vita, translated by Maureen B. Fant, foreword by Ernesto Di Renzo
43. *Cuisine and Empire: Cooking in World History,* by Rachel Laudan
44. *Inside the California Food Revolution: Thirty Years That Changed Our Culinary Consciousness,* by Joyce Goldstein, with Dore Brown
45. *Cumin, Camels, and Caravans: A Spice Odyssey,* by Gary Paul Nabhan
46. *Balancing on a Planet: The Future of Food and Agriculture,* by David A. Cleveland
47. *The Darjeeling Distinction: Labor and Justice on Fair-Trade Tea Plantations in India,* by Sarah Besky
48. *How the Other Half Ate: A History of Working-Class Meals at the Turn of the Century,* by Katherine Leonard Turner
49. *The Untold History of Ramen: How Political Crisis in Japan Spawned a Global Food Craze,* by George Solt
50. *Word of Mouth: What We Talk About When We Talk About Food,* by Priscilla Parkhurst Ferguson
51. *Inventing Baby Food: Taste, Health, and the Industrialization of the American Diet,* by Amy Bentley
52. *Secrets from the Greek Kitchen: Cooking, Skill, and Everyday Life on an Aegean Island,* by David E. Sutton
53. *Breadlines Knee-Deep in Wheat: Food Assistance in the Great Depression,* by Janet Poppendieck

54. *Tasting French Terroir: The History of an Idea,* by Thomas Parker
55. *Becoming Salmon: Aquaculture and the Domestication of a Fish,* by Marianne Elisabeth Lien
56. *Divided Spirits: Tequila, Mezcal, and the Politics of Production,* by Sarah Bowen
57. *The Weight of Obesity: Hunger and Global Health in Postwar Guatemala, by* Emily Yates-Doerr
58. *Dangerous Digestion: The Politics of American Dietary Advice,* by E. Melanie duPuis
59. *A Taste of Power: Food and American Identities,* by Katharina Vester
60. *More Than Just Food: Food Justice and Community Change,* by Garrett M. Broad
61. *Hoptopia: A World of Agriculture and Beer in Oregon's Willamette Valley,* by Peter A. Kopp
62. *A Geography of Digestion: Biotechnology and the Kellogg Cereal Enterprise,* by Nicholas Bauch